CW01649040

Financing international projects

International Construction Management Series No. 3

Financing international projects

A. D. F. Price

International Labour Office Geneva

Price, A. D. F.
Financing international projects
Geneva, International Labour Office, 1995. (International Construction Management Series No. 3.)
/Management development/,/Guide/,/Financing/,/Development project/,/Industrial project/,/Construction industry/. 11.02.4

ISBN 92-2-108747-6
ISSN 1020-0142

ILO Cataloguing in Publication Data

ILO publications can be obtained through major booksellers or ILO local offices in many countries, or direct from ILO Publications, International Labour Office, CH-1211 Geneva 22, Switzerland. A catalogue or list of new publications will be sent free of charge from the above address.

Printed in the United Kingdom EPL

PREFACE

The International Labour Office, through its Construction Management Programme, has a continuing interest in the development of construction enterprises throughout the world. Specifically, this interest stems from the dual role of construction not only as a significant source of direct employment but as a sector which contributes, through its wide range of operations and projects, to the growth and development of virtually all other economic sectors.

Many construction enterprises have chosen to limit their operations to the home market. Others appreciate the scope for diversifying into international projects, but are wary of the risks involved in venturing into an alien environment. Others still have suffered severe losses as a result of international ventures, and have resolved to avoid them in future. The caution regarding international contracting is well-founded, but the potential attractions remain. The world construction market is very large, and the projected demand for improved infrastructure and shelter suggests that it will continue to offer attractive opportunities to the discriminating bidder. Many national contractors could offer a competitive service in selected areas, provided they make realistic bids based on a proper appreciation of resources and risks.

The books in the International Construction Management (ICM) series have been written for engineers and members of other construction professions who are working in such enterprises, and who wish to fit themselves for a career in international project management. They are also suitable for undergraduate and postgraduate engineering students. Such readers are likely to have a technical background in civil engineering or a related discipline, but may have only a limited understanding of accounting, finance, strategic management, marketing and commercial topics. While there are numerous books that tackle the latter topics in a national context, the ICM series seeks to provide an integrated introduction to international construction management as a basis for making informed investment and operating decisions so as to minimize risk and improve the profitability of construction enterprises.

The ICM series is ideal for both individual and group study, since relatively complex topics are carefully explained and the active participation of the reader is ensured by the extensive use of case studies, worked examples and exercises. Although United Kingdom practice is used as a frame of reference, the books encourage the reader to compare alternative approaches and adapt the application of techniques to suit local regulatory, contractual and tax regimes.

This book, *Financing international projects*, discusses in detail the main methods of financing international construction projects, and is liberally illustrated with case studies. Some of the sources of finance discussed are usually associated with financing individual companies, while others are related to financing projects. The book thus covers the full range of financing needs related to the construction industry. Part I (Chapters 1–3) deals with the principles of project financing and common mechanisms employed by lenders. Part II (Chapters 4–6) reviews the three main sources of international project finance available to international contractors (commercial funding, development bank funding and support from their own government) and Part III (Chapters 7–10) describes some recent trends in alternative approaches to financing and spreading the risks inherent in large projects.

Chapter 1 has been designed to provide the reader with an appreciation of the principles underlying the provision of corporate and project finance. Chapter 2 describes the growth of domestic, international and multinational banking, as well as the major activities of international banks and how they assess international lending risks. Chapter 3 outlines some common practices and procedures used by commercial banks in putting together syndicated loans for large projects, including a background to Eurocurrency lending.

Chapter 4 discusses project financing from commercial sources, emphasizing the importance of adequate owner/sponsor or third party guarantees. Chapter 5 outlines the specific project financing procedures operated by the main development banks, including a description of the activities to be undertaken at various stages of the project cycle. Chapter 6 describes some of the ways in which governments seek to back the project export initiatives of their national contracting companies through mechanisms such as soft loans, preferential loans and export credits.

Chapter 7 explains how the risks inherent in large projects can be shared through joint venture agreements, emphasizing the need for adequate documentation and a clear management structure taking into account possible differences in business philosophy among the parent companies. Chapter 8 discusses the recent trend towards barter arrangements and countertrade to finance projects in countries where conventional project finance is difficult to mobilize. Chapter 9 offers an introduction to another innovative approach, Build-Operate-Transfer (BOT), in which the contractor takes responsibility for managing the completed facility and recovers all or part of the contract price through revenue earned from its operation during a concessionary period. Chapter 10 concludes the book with a basic guide to insurance and performance bonds.

This book was prepared by Andrew Price, Senior Lecturer in the Department of Civil Engineering of the Loughborough University of Technology, United Kingdom, and edited by Derek Miles, Director of Overseas Activities in the Department of Civil Engineering of the Loughborough University of Technology, and previously Coordinator of the ILO Construction Management Programme.

ACKNOWLEDGEMENTS

This book in the International Construction Management (ICM) series was designed to assist the China International Contractors' Association (CHINCA) to meet the needs of its member corporations in building the broad management expertise needed to expand their operations on the international market. It was prepared within a technical cooperation project carried out by the ILO as executing agency under the United Nations Development Programme (UNDP). In view of the dearth of textbooks and training material that would be suitable for the needs of senior and middle-level management staff of these enterprises, it was decided to commission a series of texts that would enable them to improve their understanding of the range of specialist management skills that are required to compete effectively in the international construction market.

We are grateful to the staff of CHINCA and its member corporations for their advice and assistance during the field testing of the material upon which this book is based, and the agreement of both CHINCA and UNDP that it should be made accessible to a wider audience.

CONTENTS

Figures

Tables

PART I

PRINCIPLES OF PROJECT FINANCING

SOURCES OF FINANCE

1

1.1 Introduction

The construction industry, unlike the manufacturing and service industries, tends not to operate from a fixed location but moves from site to site wherever work is available. A contractor's principal physical asset, equipment which also moves from site to site, is subject to rough treatment and has a useful life of just a few years. These assets provide a narrow collateral base. Therefore, when assets are examined together with other problems, such as one-sided contracts or delays in payment, the argument for financing construction projects is substantially weakened. Most international construction projects face additional problems, for example fluctuating exchange rates or import restrictions. These additional risks make the raising of finance for international projects a difficult and complex process.

This book examines in detail several aspects of financing international construction projects, starting with a review of the various sources of finance available to the construction industry. Some of these relate to sources of finance available to individual companies (e.g. bank overdrafts), while others are more specifically concerned with the financing of individual projects (e.g. export credits).

1.2 Financing construction companies

When most construction companies are created, the initial sources of finance are usually provided by private individuals and ploughed-back profits. Most other investors are generally not prepared to take the risks associated with an unknown enterprise without some form of guarantee. As the company grows and takes on larger contracts, these initial sources become insufficient to fund the required heavy capital investments which are often beyond the means of an individual or a company. A need for finance from external sources arises, and there are many different approaches that can be taken, so long as the company has gained the confidence of the moneylenders through a proven track record. The types of available finance can be separated into three groups: short-term finance for durations up to one year; medium-term finance for durations of between one and five years; and long-term finance for durations of more than five years.

Short- and medium-term finance can be used to overcome immediate cash-flow problems. Materials have to be purchased, plant hired, workers and subcontractors paid, and so on, before payment is received for the finished product. To ease these cash-flow problems, construction companies have at their disposal many short- and medium-term sources of finance, some of which are discussed in sections 1.3 and 1.4.

Long-term finance may be required to purchase equipment and plant, or simply to help the company expand. In some cases, medium-term finance that is repeatedly renewed when the term expires becomes long-term finance. Lenders will always try to assess the risks involved in financing companies or projects. The higher the risk, the more expensive the finance. Long-term finance is considered to be risky because of the time-scale involved. Short- and medium-term finance is less risky and therefore tends to be cheaper. Naturally, the firm must be well managed and profitable before the lending institutions consider any application for finance.

1.3 Short-term finance

1.3.1 Working capital

The working capital of a company at any time can be defined as the sum by which current assets exceed current liabilities. The current assets of a company are shown on the balance sheet and consist of stock held, cash in the bank, debtors and short-term investments. The current liabilities of a company are also shown on the balance sheet and represent the amount of money the company must pay out within one year; this includes creditors, tax payments, dividends to the shareholders, repayments on short-term loans and bank overdrafts. Thus, working capital is the amount of money a company has at its immediate disposal. It is also viewed as the "lubricant" of the daily transactions needed to run the business.

When a company is to maximize its profits, it must ensure that the ratio of current liabilities and current assets is well balanced. When current liabilities exceed current assets, then the company is "out of lubricant" and near to failure. In order to meet its daily payments, it will have to request longer to pay, borrow money, or sell some of its fixed assets. All three solutions are far from ideal: the first damages the company's reputation, the second increases costs and the third reduces overall operating capacity. When current assets exceed current liabilities, or in other words the company's working capital is excessive, the company is not utilizing its assets to the full, and it is sensible to make a positive use of surplus working capital by investing. However, care must be taken to ensure that a healthy degree of liquidity is maintained.

Delaying payments of trade debts is one way of increasing a company's working capital over the short term. Many contractors use this approach and run the risk of being blacklisted by their suppliers.

1.3.2 Bank overdrafts

Banks are the most widely used source of finance, supplying money by means of overdrafts or loan facilities. Merchant banks tend to provide the larger loans and hence play an important role in the financing of construction projects, whereas the clearing banks provide smaller loans and overdraft facilities.

Overdraft facilities are offered mainly by the clearing banks and have the advantage of being flexible in that money can be drawn against the facility up to a previously agreed limit, as the need arises. Interest is calculated on a daily basis on the money outstanding and usually varies between 2 and 3 per cent above the base rate. The credit limit is set in relation to the client's creditworthiness, the nature of the project, the expected cash generation during or on completion of the project, the management team's experience, project risk, and various other factors which vary from bank to bank. Overdraft facilities are considered to be short-term finance, and the contractor or the client is usually granted up to one year to balance the account.

However, as overdrafts are "repayable on demand", the bank can demand a settlement at any point in time. If the borrower fails to repay the debts, the bank can take a number of measures:

- When the overdraft is secured, collateral deposited with the bank can be sold to recover the money. Collateral could be the land on which the project is being built.
- When the overdraft is unsecured, the client can be sued for bankruptcy. However, this is not advisable for the bank, as the borrower may not have enough cash to cover the debts and the bank would lose its only real security, which is the successful completion of the project.
- A new solution could be negotiated with the borrower in order to improve the situation and ultimately reach the completion of the project, when the bank could recoup all its money and interest.

The first two measures should be taken only as a last resort, as it is usually more profitable for both the bank and the client to complete the project.

1.3.3 Trade credit

According to Christy and Roden (1973, p. 21), "Trade credit arises when a business receives goods or services from a supplier without the requirement of an immediate cash payment". It is a very specific source of finance as it does not provide cash on credit, but goods and services. The amount of trade credit provided to a company depends on the company's creditworthiness and how important the supplier considers the company to be. Trade credit comprises three elements: the discount available on the invoice price; the discount period starting from the day when the invoice is received; and the time when the gross amount of the invoice has to be paid.

If trade credit is to work, suppliers must insure themselves against bad debts. Insurance companies are prepared to take some of the risk, but will impose their own terms and conditions relating to the amount of goods or services to be advanced to customers. This introduces the notion of the rating of firms, whereby suppliers rate a company based on its past credit record, its reputation and its importance as a customer to the supplier.

Companies subject to trade credit should ensure prompt repayment at the set dates, as any delay could pull their rating down. Once a company becomes a regular customer, it can expect to have its credit limit increased by its suppliers.

Trade credit is widely used, mainly by small and medium-sized firms which rely heavily on this source of finance. It is considered by many companies as interest-free short-term lending, although strictly speaking it is not interest free. This is because most suppliers offer a discount if immediate payment is made. The main characteristics that make trade credit an attractive option to construction companies are as follows:

- it is readily accessible and does not jeopardize the firm's line of credit with the bank;
- it can be used to smooth out short-term cash-flow problems experienced by many small and medium-sized companies; and
- it is relatively inexpensive.

1.3.4 Factoring

During the current recession, some of the larger construction companies have abused trade credit by excessively delaying payments, which has pushed many smaller companies to the brink of receivership. Consequently, factoring and invoice discounting have become important countermeasures for the supplier. When sales are made on credit terms, suppliers have to wait several weeks before being paid by their customers. The delay in payment allowed will depend on the supplier's policy and the creditworthiness of the customer. Since the customer is being financed by the supplier, the concept of factoring is to delegate the collection of money to a factoring company.

The two main types of factoring are "service factoring" and "undisclosed factoring". In the case of service factoring, the factoring company reviews the debt and the creditworthiness of the customers and pays the supplier, at the end of each month, the amount owed by its clients minus a percentage fee. The fee charged by the factoring company depends largely on how much risk is involved, the supplier's turnover and the credit period. The factoring company takes on the responsibility of collecting the debt, and consequently invoices the customers directly, as illustrated in figure 1.1. The advantages of factoring are that much-needed working capital can be quickly turned from debts into cash, and the factoring company can be more aggressive with bad debtors than can the supplier. The main disadvantage of factoring is obviously the percentage fee charged to the supplier, which has to be passed on to the customers.

Figure 1.1. Service factoring

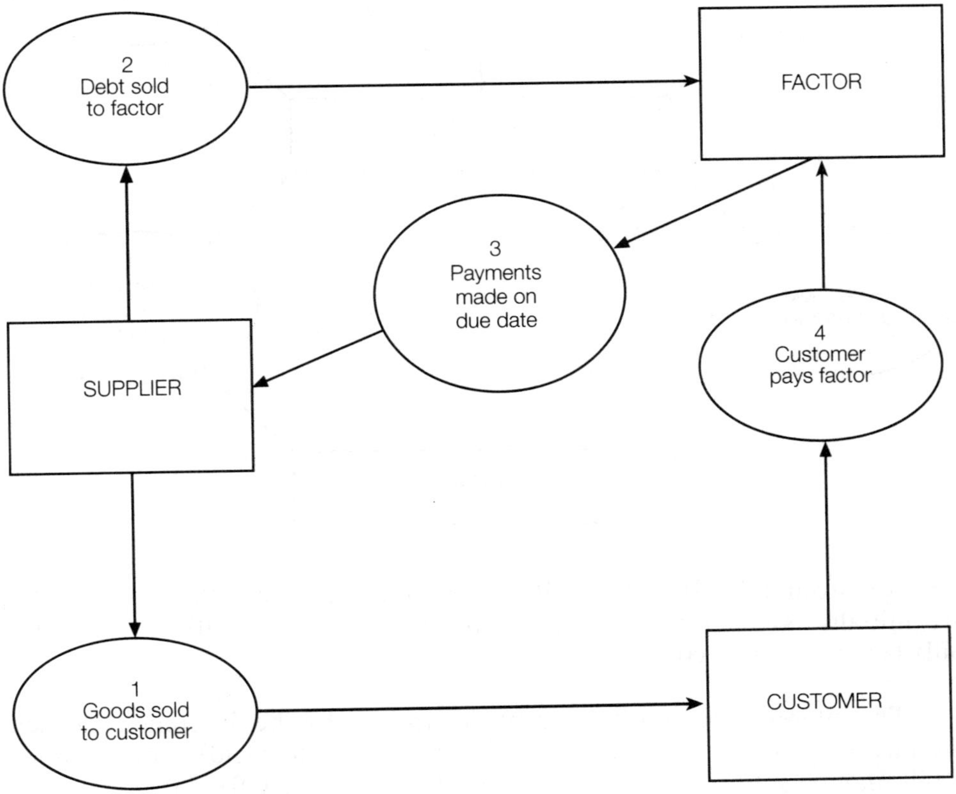

Some suppliers wish to avoid the intervention of the factoring company in the supplier/customer relationship. Undisclosed factoring, as illustrated in figure 1.2, can help to alleviate this fear. The factoring company purchases the debt before goods are delivered to the customer and appoints the supplier as an agent responsible for collecting payment from the customer.

1.3.5 Bills of exchange

Bills of exchange are effectively time drafts drawn on one company or trader by another to finance the import or export, or sale of goods. The face value of most bills varies from a few thousand pounds to an upper

Figure 1.2. Undisclosed factoring

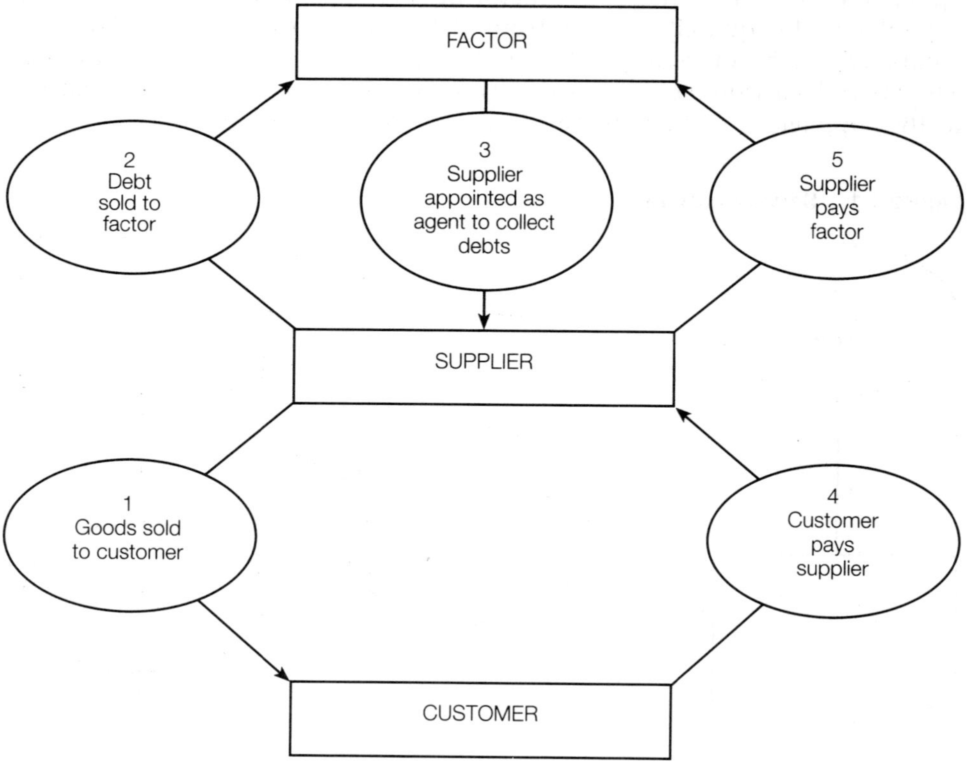

limit of around US$1 million. Bills are short-term debt instruments and are initially issued for a period of one to six months, with three-month bills being the most common.

The interest rate of a bill can be determined by the difference between the face value and the selling value of the bill (i.e. the yield). For example, a bill may be sold for US$50,000 with a face value of US$55,000 maturing after a period of six months. The yield is then US$5,000 and the interest rate is equivalent to 20 per cent per annum.

There are two main types of bill, known as "trade bills" and "bank bills", as illustrated in figure 1.3. Any company can issue a trade bill. The bill is sent to the buyer or drawee who accepts it by signing and returning it to the seller or issuer. Trade bills are payable to the bearer. The issuer can then keep it to maturity, or sell it (discount it) at a slightly lower price than the face value. In the case of a bank bill, an "acceptance house" intervenes in the process. A merchant bank, acting as an acceptance house, can create short-term marketable assets by approving commercial bills. If the financial institution acting as the acceptance house is satisfied with the buyer, a letter of credit will be issued to the buyer.

Figure 1.3. Bills of exchange

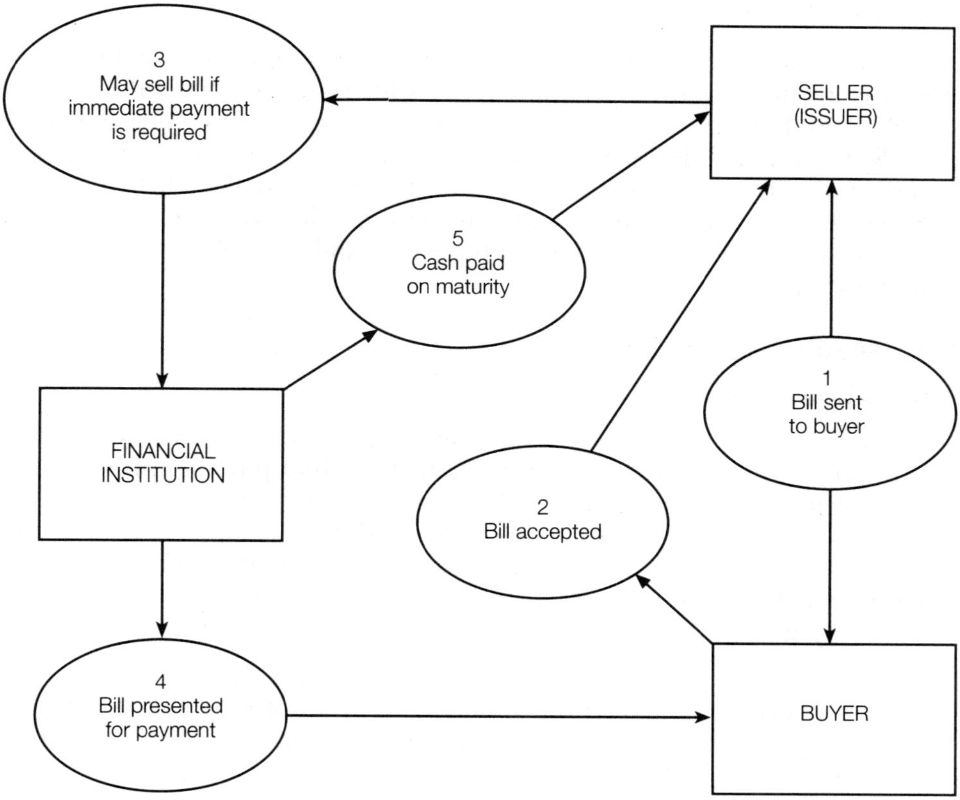

1.3.6 Commercial papers

Commercial papers are a form of short-term borrowing facility with maturities from seven to 364 days. Companies can thus obtain unsecured finance directly from investors, bypassing banks and bond markets, although banks are often used as agents to place the paper. This source of finance has been used for a long time in the United States. However, the Sterling Commercial Paper (SCP) market in the United Kingdom is fairly new. It opened in May 1986, shortly after Her Majesty's Treasury and the Bank of England introduced new regulations permitting the issue of sterling-denominated commercial papers. Since then, it has become established as a competitive money market instrument for a wide range of borrowers and investors in the United Kingdom. It takes the form of a promissory note denominated in sterling and payable to the bearer. The note carries the words "Negotiable Bearer Sterling Commercial Paper".

SCP may be issued in denominations of £500,000 or £1 million. Issues are usually on a non-interest-bearing basis at a discount to the face value, although this is not always the case and it is possible for the SCP to bear a specified interest rate. The SCP market is accessible to both British and overseas-based companies, but they must have ordinary or preference shares listed on the International Stock Exchange in London and net assets of at least £50 million.

There are several approaches to issuing SCP, the most common of which is an uncommitted programme by which a dealer buys the bulk of the SCP and distributes it to investors. Alternatively, a company may issue SCP directly to investors or through a panel of banks on a competitive tender basis. The issue of SCP is made through a bank, which becomes responsible for authenticating and safeguarding the notes. The bank also controls the payment flows between the issuer and the holder both at issue and maturity.

1.4 Medium-term finance

1.4.1 Finance houses

Finance houses operate by extending funds to the industrial, commercial and consumer sectors in the form of hire purchase, leasing and loan agreements. Like commercial banks, they obtain capital by accepting deposits from customers, including industrial and commercial concerns. They also borrow from banks and other financial institutions. Finance houses have to adjust to the various credit and ceiling controls that are imposed by government.

During the 1950s, finance houses were mainly concerned with providing consumer finance to facilitate credit sales of goods. In the 1960s, they started to offer leasing arrangements which soon became an important source of finance to both industry and the public. Under an industrial hire-purchase arrangement, a contractor purchasing plant usually pays an initial cash deposit to the supplier, and a finance house provides the outstanding balance. The contractor reimburses the finance house by means of regular instalments over a specified time period, the amount and number of instalments reflecting the interest rate charged by the finance house. Although finance houses appear to dominate the hire-purchase market, many suppliers have the necessary legal and financial expertise to provide hire-purchase facilities from their own resources.

One advantage of hire purchase agreements is that companies can still obtain this facility even if they have no access to a bank overdraft, or if their overdraft limit has been reached. However, some hire purchase deals are a very expensive source of finance. Another advantage of hire purchase is that the equipment is the property of the contractor, although the finance house will have some claim on it if the contractor defaults on the instalments.

In the case of leasing arrangements, the finance house buys the equipment from the supplier and leases it to the contractor for monthly instalments over a pre-agreed period of time (usually two to three years). The advantage of this method is that the contractor is not required to produce a cash deposit, and this helps to improve the cash-flow situation.

The main drawback is that the contractor does not own the equipment at the end of the leasing period, but can buy it from the finance house. There are also tax benefits to be associated with the leasing of equipment which, as an off-balance sheet debt, does not appear as a debt in the company's annual accounts.

1.4.2 Sale and leaseback

When a company gets into financial difficulties or requires additional capital in order to finance future growth, it may be necessary for it to sell off some assets. If the company owns freehold property or has a long leasehold, capital can be raised by mortgaging or selling the property at the market price. However, mortgages can usually only be obtained for a proportion of the market price and the sale implies that there will be a new owner/occupier. A sale and leaseback arrangement is one way for the company to get around this problem and still raise as much money as possible against its property.

The sale and lease are arranged concurrently in order to ensure that the existing owners continue in occupation. The new owners obtain a return on their investment through the rental charged, which is based on the sale price. It is therefore possible to raise more than the market price on the sale of the property, as long as the occupants are prepared to pay a relatively high rental. The sale and leaseback arrangement can be used by companies to convert their properties into cash, thus increasing their working capital and reducing cash-flow problems.

Investors have to consider the long-term situation and must be prepared to have their capital tied up for considerable periods. Sale and leaseback arrangements are normally made with insurance companies, but some financial trusts, property investment companies and pension funds are also prepared to provide this sort of finance.

1.4.3 Bank loans

Bank loans are considered to be medium-term finance and are obtained for projects requiring finance for between one and five years. Interest rates are fixed on issue of the loan and are usually higher than rates for overdrafts, resulting from the increased risk to the bank because loans are not repayable on demand. The time and credit limits imposed on bank loans can be renegotiated and extended if the project encounters some delays or difficulties. Interest is usually paid on a monthly basis on the total sum issued regardless of the decreasing balance. In addition to interest, banks often charge a set-up fee which can amount to 0.5 per cent of the sum lent.

Although more expensive than an overdraft, loans provide the client with the certainty of finance over a medium-term period and inspire the confidence to plan ahead. Sometimes a client may require a five-year loan, but the bank can provide only a one-year loan because of its own lending limit. If this is the case, the bank could provide a roll-over loan whereby the loan is renewed every year but the interest rate is fixed at the beginning of each year; thus, the client is taking the risk of fluctuating interest rates. If the client is not satisfied with the deal offered by the bank, or if the bank has turned an application down, it does not necessarily mean that the project is not worth financing. The bank could have reached its lending limit for the construction sector. It is therefore important that the client "shops around" in order to get the best deal on offer. A good reputation and a strong balance sheet always help the client to obtain better terms of finance from the banks.

1.5 Long-term finance

The main sources of long-term finance are retained profits, share issues and debentures. Therefore, long-term finance is usually secured against the company's assets after all other debts have been paid. These sources are discussed below.

1.5.1 Retained profits

One of the most significant sources of capital for the construction industry is ploughed-back profits or profits retained by the business. This source of finance is attractive because it does not require the payment of any interest on the money raised. At the end of the tax year, a proportion of profit is paid to the government in the form of corporation tax, and to the shareholders as dividends. The remainder can be reïnvested in the business for future expansion or to fund new projects. Profits retained in the business are shown on the balance sheet as reserves, and contribute by increasing the working capital.

The Board of Directors has the difficult task of deciding how much profit should be paid out in the form of dividends and how much should be set aside for future investment. It is a fine balancing act between keeping the shareholders satisfied with the return on their investment, and at the same time ensuring that enough funds are made available to maintain and improve the company's financial well-being. If dividends are too low, shareholders may well sell their stocks and push down the price of the company's shares. If there is insufficient reinvestment, then the company may lose its competitive edge. Most companies will have a corporate policy or strategy which influences how much profit is paid out as dividends and how much is retained.

1.5.2 Share issues

All companies use a combination of sources to finance the purchase of assets and the day-to-day running of the business. These sources usually comprise share capital, long-term loans and short-term facilities. Substantial share capital is required to encourage external sources to provide additional finance. There are two types of shares, ordinary and preference, and these are defined by Rutherford (1992, p. 417) as:

> . . . a portion of the financial capital of a limited company which gives the holder an entitlement to a fixed return, in the case of preference shares, or, in the case of ordinary shares, a variable dividend decided by the board of directors. Ordinary shares are also known as equities.

Ordinary shareholders are the owners of the company, and carry full voting and dividend rights. However, dividends are paid out to ordinary shareholders only after payments have been made to the holders of debentures (section 1.5.3), loan stock and preference shares. Preference shareholders do not have voting rights, but they do have preferential rights for dividends, usually in the form of a fixed dividend. They rank just before ordinary shareholders but after debentures in respect to dividend payments. Obviously, these shares with greater preference are subject to less risk than ordinary shares, which is reflected in the lower rate of return. There are three types of preference shares, as explained below:

- cumulative preference shares guarantee the holder a cumulative dividend if the company had been unable to pay previous years' preference dividends owing to financial difficulties;
- participating preference shares entitle the holder to a fixed amount of interest, or a dividend and extra earnings above a pre-set level; and
- redeemable preference shares are fixed-term stock redeemable at an agreed date and price, set by either the company or the shareholders.

Companies can raise additional finance by issuing new shares either to the general public or its own shareholders. The process of issuing new shares can be expensive and is therefore restricted to raising large amounts of capital which can be injected into the company in order to finance new projects. The advantage of raising additional finance through a share issue is that the company does not have to pay a fixed interest rate on the money raised. However, it will have to pay out extra dividends from the profits it makes. The main disadvantage is that new voting rights may be introduced which could shift the existing balance of control.

1.5.3 Debentures

The term debenture can be confusing. In the United States a debenture signifies an unsecured debt; however, in the United Kingdom it usually refers to secured stock. In both cases, a debenture is a written acknowledgement of a long-term debt with maturities of more than 15 years. If shorter maturity periods are required, notes are issued. Pizzey (1990, p. 334) defines debenture finance as follows:

> A debenture is the written acknowledgement of a loan. When this acknowledgement is made by a company, it is a contract made under the company seal, providing for a fixed rate of interest to be paid on the sum loaned to the company and specifying

terms for repayment of the principal at the end of the period. The debenture, therefore, is a long-term finance.

Debenture holders are not members of the company, but creditors. This means that, unlike shares, a fixed rate of interest is charged on the debt whether or not the company is making a profit. Debentures are a cheaper way to raise finance than shares because of their tax advantage. The interest a company pays on a debenture is free of tax, unlike a dividend payment which is made from the taxed profits. However, even when times are hard, interest on debenture loans must be paid whereas dividends need not be.

A debenture is much less risky to the investor than ordinary or preference shares because of the certainty of interest. Debenture holders also have priority rights over all the company's assets in the case of liquidation. In addition, the debenture can be sold on the market, and the price fluctuates to reflect current interest rates. From the company's point of view, debenture finance is attractive because debenture holders do not possess voting rights; hence there is no dilution of controls when new issues are made.

The convertible option is a special form of debenture and is increasing in favour in the United Kingdom, particularly with larger companies. This is a fixed-interest loan which carries with it the right to convert into equity shares on fixed terms at some future date.

1.6 Project financing

There are many different approaches to project financing. This makes it difficult to have one definition that covers all the variations that are available. However, Howcroft (1986, p. 2) gives the following definition that can be applied to most situations: "Project finance: A financing of a particular economic unit under which the lender partially or wholly relies upon the assets of the unit for collateral and upon the cash flow from the unit as his source of funds for repayment of debt."

To enable finance to be secured for any project, an initial project appraisal will have to identify both the most probable project cost and the potential viability of the project. The potential sources of financing international construction projects should be considered in relation to the following:

- project objectives;
- finance costs;
- sources of plant, equipment, etc.;
- currencies available; and
- export earnings from the project.

The initial search for finance should investigate the suitability and availability of the following sources:

- project sponsors and partners;
- development banks;
- export credit finance;
- domestic banks;
- Eurocurrency loan banks;
- fixed-interest capital market;
- leasing; and
- finished product financing.

The starting-point should be with the project sponsors and partners in order to identify available sources of finance. Export credit finance may be an option, depending upon the availability of plant and equipment. Potential export earnings could be used to minimize both costs and the foreign exchange risks involved.

In effect, the sponsors and partners become the shareholders. These may involve wealthy individuals, governments or suppliers with an interest in the project's success. The equity raised plays an important role in attracting sources from elsewhere. In some cases, such as the Channel Tunnel, these investors may have to accept that returns on their investments may not be available in the short or even medium term. When considering investment in foreign projects, a company should take account of the following:

- What will the rate of return be?
- How much finance can be raised in the host country?
- How much finance should be provided by the company?

The first question should be answered by the initial project appraisal. The factors influencing the need for investments from foreign partners are presented in figure 1.4, and the factors influencing the amounts contributed by equity and subordinate debt in figure 1.5.

1.7 Summary

A well-established business maxim is that short-term requirements should be financed from short-term sources of capital, while fixed assets and permanent working capital should be financed from long-term capital. The ratio of total borrowing to total capital and reserves is known as the borrowing factor. The ratio of fixed interest capital to equity capital is known as the gearing ratio. Fixed interest capital can be used to provide additional earnings on equity share capital. However, fixed interest has to be paid, even when profits are low, and this could eventually eat into equity share capital during long periods of recession. Consequently, if a company is subject to large fluctuations in earnings, a low gearing arrangement should be sought whilst ensuring that there is sufficient short-term capital to meet the company's short-term debts.

Figure 1.4. Factors influencing foreign partner investment

Equity	Debt
Type of investment grant	Cost
Location of investment grants	Restrictions
Ownership policy	Guarantees
	Currency fluctuations
	Exchange controls
	Company gearing

Figure 1.5. Factors influencing amounts contributed by equity and subordinate debt

Equity	Subordinate debt
Legal considerations	Current account balance
Permanence of capital	Permanence of finance
Taxation	Taxation
Exchange controls	Exchange controls
Lending institution	Cash requirement

References

Blommeart, A. M. M., et al. 1991. *Financial decision making* (New York, Prentice Hall).

Brearley, R. A.; Myers, S. C. 1991. *Principles of corporate finance* (New York, McGraw-Hill, 4th ed.).

Brownlie, S. M.; Harris, F. C. 1989. "A review of finance for large-scale construction", in *Construction Management and Economics* (London), Vol. 5, No. 2, pp. 115-121.

Christy, G. A.; Roden, P. F. 1973. *Finance: Environment and decisions* (San Francisco, California, Canfield Press).

Davies, P. R. 1990. *Introduction to corporate finance* (London, Chartered Institute of Bankers).

Dictionary of modern economics. 1983. (New York, McGraw-Hill, 3rd ed.).

Durury, C. 1989. "Lease financing", in *Managerial Finance* (Bradford, MCB University Press), Vol. 15, 1/2.

Dyer, L. S. 1980. *A practical approach to bank lending* (London, Institute of Bankers, 2nd ed.).

Freear, J. 1973. *Financing decisions in business* (London, Haymarket Publishing).

Harris, F. C.; McCaffer, R. 1989. *Modern construction management* (London, BSP Professional, 3rd ed.).

Howcroft, B. 1986. *Project finance* (Loughborough, Leicestershire, Loughborough University Banking Centre).

Johnson, G. T.; Hoel, L. A. 1987. "Review of financing options for highways and transit", in *Journal of Transportation Engineering* (New York, American Society of Civil Engineers), Vol. 113, No. 1, Jan., pp. 72-83.

Kemp, L. J. 1981. *A guide to world money and capital markets* (New York, McGraw-Hill).

Lee, T. A. 1976. *Company financial reporting, issues and analysis* (London, Nelson).

Midgley, K.; Burn, R. G. 1979. *Business finance and the capital markets* (London, Macmillan).

Mitchell, C. 1988. *The Sterling Commercial Paper market* (New York, Woodhead-Faulkner).

Paish, F. W.; Briston, R. J. 1982. *Business finance* (London, Pitman, 6th ed.).

Pearce, D. W. 1986. *Macmillan dictionary of modern economics* (London, Macmillan, 3rd ed.).

Peters, G. H. (1971) *Private and public finance* (London, Fontana/Collins).

Pizzey, A. 1990. *Accounting and finance: A firm foundation* (London, Cassell, 3rd ed.).

Quinn, H. 1989. "Construction: Strong foundations", in *Asian Finance* (Hong Kong), Vol. 15, No. 12, 15 Dec., pp. 22-23.

Rector, K. 1989. "Avoiding pitfalls with major construction financing", in *Texas Banking* (Austin, Texas), Vol. 78, No. 6, June, pp. 10-12.

Rutherford, D. 1992. *Dictionary of economics* (London, Routledge).

Shaw, E. R. 1984. *The London money market* (London, Heinemann, 3rd ed.).

Wilkinson, R. 1988. "Construction financing: Plan now before the crunch", in *Trustee* (London), Vol. 41, No. 2, Feb., pp. 10-11, 26.

MECHANISMS OF INTERNATIONAL BANKING TRANSACTIONS

2

2.1 Introduction

Documentary evidence of international banking dates back as far as the twenty-first century BC, where letters of credit and bills of exchange were used as a means of international lending. More recently, around the fourteenth and fifteenth centuries, the Bardi, Peruzzi and Medici Florentine banking houses had branches and subsidiaries throughout Europe built upon the growth of international trade in wool and silks. During the nineteenth century, the economic development of the United States was financed by capital provided by British and European banks. In the early part of the twentieth century, the United States became a major source of net capital for Latin America, Canada and Europe (especially Germany). Modern international banking probably started during the 1920s, when banks in London took deposits from resident and non-resident customers in US dollars and other foreign currencies. In addition to this, banks in Vienna and Berlin accepted US dollars and sterling deposits. The next step was a natural progression from banks providing loans in the currency of their operation to banks providing loans in a wide variety of currencies.

Modern international banking activities are generally undertaken by a few key international banks. This chapter explains the difference between international banking and international bonds, distinguishes between the various types of banking operation and the types of bond used on the capital market, and covers the major activities of international banks. Loan documentation, international lending risks, transactions, and forward buying and selling are also covered. Finally, the chapter discusses the cost of financing in foreign exchange.

2.2 Banking markets and capital markets

Most countries have resorted to external borrowing from the international market in order to finance industrialization and development programmes, or accommodate fluctuations in the world economy which influence their international trading position. There are two main approaches to the financing of these programmes, namely on the banking market, or on the capital market.

The main feature of banking is that the bank places itself between the lender and the borrower. The bank obtains funds from its investors and lends them to the borrower at a higher rate of interest. However, the issuing of bonds on the capital market enables finance to move from surplus regions to deficit regions without the involvement of banks as intermediaries. The activities on the banking and capital markets can be classified as follows in relation to type of currency used and location:

- Capital market: Domestic bonds
International bonds
- Banking market: Domestic banking
International banking
Multinational banking

2.3 Domestic, international and multinational banking

The three main banking activities can be classified as domestic, international and multinational depending upon the following characteristics:

- currency denomination in terms of domestic currencies (onshore or offshore) and foreign currencies (Euromarket);
- location of operation (home or abroad);
- ownership of bank (home ownership or foreign ownership); and
- location of customer (resident home or abroad) in relation to location of operation.

Domestic banking has traditionally been viewed as transactions between banks based in country A and residents of country A in the currency of country A. Transactions of non-residents based in country A may not form part of international business, and can also be deemed as domestic banking when transactions are in the currency of A by a foreign bank.

International banking refers to the cross-currency and cross-border activities of banking. These can include the traditional activities of taking deposits from non-residents (foreigners) or lending to non-residents in the currency of the country in which the bank is based. International banking activities can be separated into traditional foreign banking operations and Eurocurrency banking operations.

Eurocurrency banking involves the borrowing or lending in currencies other than the domestic currency of the country in which the bank is located. Large sums of money are usually involved and the main currency traded is US dollars, hence the term “Eurodollar”.

Traditional foreign banking operations are mainly domestic currency transactions by indigenous banks with non-residents (either banks or non-banks) based overseas. In addition, banks and businesses maintaining foreign currency balances with foreign banks also constitute traditional foreign banking.

Multinational banking was defined by Robinson (1972, p. 4) as "operating a bank in, and conducting banking operations that derive from, many different countries and national systems". This broad definition encapsulates all banking operations including traditional foreign banking, Eurocurrency banking, domestic wholesale and domestic retail operations. However, Lewis and Davis (1987, p. 219) focus on the global nature of the bank's location and ownership, and more narrowly define multinational banking as "the location and ownership of banking facilities in a large number of countries and geographic regions".

In addition to classifying banking operations in terms of domestic, international and multinational, it is important to consider the scale of operation in terms of retail banking and wholesale banking.

Retail banking involves the traditional banking services provided to the general public (i.e. consumers, small businesses and commercial enterprises). Retail banking does not include any Eurocurrency activities.

Wholesale banking usually involves the borrowing or lending of large sums of money by the major banks. It can take the form of providing bank services, loans or loan securities to large corporate organizations. A large proportion of wholesale banking involves interbank activities or dealings with other financial institutions. Eurocurrency business is characterized by the features associated with wholesale banking.

2.4 Domestic, international and foreign bonds, and Eurobonds

Bonds are issued on the capital market and take the form of an interest-bearing certificate stating the existence of debt and the amount owing to the holder of the document. The certificate must be redeemed to obtain repayment of the debt. The capital market is largely unregulated, and capital can flow freely between regions without involving financial intermediaries such as banks and other institutions. However, banks often organize and underwrite the exchange process. The security (or bond) is issued by the borrower, and the lender has no claim on the bank. Bonds can be classified as domestic, international and foreign bonds, and Eurobonds. **Domestic bonds** are issued by borrowers in the country where they reside. **International bonds** are issued by borrowers in countries other than those in which they reside and can take the form of foreign bonds or Eurobonds. **Foreign bonds** are issued by non-resident borrowers on the market of a single country in that country's currency. However, **Eurobonds** are sold in several countries simultaneously. Most foreign and Eurobond issues are underwritten and sold by international bank syndicates or issuing houses.

2.5 Major activities of international banks

There are many international lending markets, with the Eurocurrencies being a major market in their own right. Although most banks will handle some degree of international business for their customers, there are only a few purely international banks. International lending excludes retail or personal lending, but not necessarily interbank lending. It includes acceptances (i.e. an accepted bill of exchange signed by a bank), letters of credit (i.e. documents issued by a bank on behalf of a buyer on another bank or itself), bid bonds, performance guarantees and all other types of finance, so long as they satisfy at least one of the following criteria:

- the currency is foreign to either the borrower or the lender, or both;
- it is used to finance cross-border investment or international investment;
- it is provided by a syndicate of banks of more than one nationality; or
- the guarantor is based outside the country in which the loan is made.

International banks perform a wide variety of activities, some of which are listed below:

- receiving deposits and making loans in a domestic or foreign currency to foreign governments, enterprises and individuals;
- managing or acting as agents for syndicated loans (see Chapter 3);
- dealing in gold and precious metals, international money transfers and other aspects of foreign exchange transactions;
- providing letters of credit, standby letters of credit (i.e. available as required), multiple currency credit lines, bank acceptances and Euronote assurance facilities;
- trading in currency futures and options, financial futures and options, and interest rate and asset swaps (see section 2.8); and
- distribution of Eurocommercial paper (i.e. commercial paper issued in a Eurocurrency), underwriting and placement of Eurobond issues.

2.6 Loan documentation

Full documentation of short-term facilities is not usually required but must be legally valid. The documentation of medium-term agreements is an integral part of the negotiation process and forms an essential record of the agreed terms and conditions of a loan. The agreement must be able to stand up in court, although the aim is obviously to avoid having to go to court.

Prior to the 1960s, the main providers of medium- to long-term international loans were institutions rather than banks. As most of these institutions were based in New York, loan documentation was usually based on New York law and practice. More recently, a Euromarket approach to loan agreements has emerged, with some standard clauses creating a common format to agreements. Some of these standard clauses can be used for new types of transaction, but others may not be appropriate and will have to be replaced. Hence, a well-established foundation of standard practice is available when producing agreements for very new forms of financing, the most common element being that English or New York law should be used to draw up the agreement. Occasionally other laws may be preferable, but the banks must be given a very good reason if they are to depart from the laws that they are familiar with.

Some of the specific clauses that may be used to form the documentation for a medium-term loan are listed below:

- changes in circumstances;
- the Eurodollar disaster clause;
- unrepresentative cost;
- sovereign immunity;
- governing law and jurisdiction;
- cross default;
- material adverse charge;
- notes (e.g. promissory notes);
- financial covenants and ratios; and
- negative pledge.

A more comprehensive explanation of these clauses can be found in Donaldson and Donaldson, 1982.

2.7 International lending risks

Banks may lend without any analysis on security but on the strength of the borrower's name alone. However, this is not a very sound approach and few banks admit to it. There are several analytical approaches, but these can be divided into two categories as follows:

- *The liquidation method*

  In this approach, the bank assumes that the company has to be sold and adjusts the value of the company's assets and liabilities accordingly. A loan may be made if the bank is satisfied that the assets clearly exceed the liabilities.

- *The going concern method*

  The bank concentrates mainly upon the company's balance sheet in order to satisfy itself that the proposed level of operations can be supported. Sensible projections are made with regard to the future level of operation, and the bank will make a loan so long as the generated cash flows are sufficient to repay the debt.

There are many different types of risk associated with domestic lending. These include:

- liquidity risk;
- interest rate risk;
- credit or asset risk; and
- contingent liability risk.

International banks not only face the same risks but have to deal with additional risks, especially with cross-border and cross-currency transactions. Although the basic principles of country and company analysis are similar, the size and complexity of a country make the analysis considerably more complicated. The Bank of International Settlements (BIS) is provided with detailed accounts of countries borrowing from the banking system. The additional risks associated with international lending include:

- currency risk; and
- regulatory risk.

Banks have to set limits on the degree of exposure within individual countries in order to avoid overexposure. The following risks also have to be considered in addition to those already mentioned:

- political risk;
- political/economic risk;
- economic/commercial risk; and
- sovereign risk

2.8 Transactions, forward buying and selling

Eurocurrency activities usually involve loans and repayments. A loan made in Eurodollars can be claimed against "ordinary" dollars, the cost of the loan being the interest rate charged. However, foreign exchange dealings are very different and involve mainly purchases and sales. The cost of dealing is charged through the exchange rates agreed. There are two types of transaction involved in the foreign exchange market, namely:

- spot transactions; and
- forward buying and selling.

When foreign currencies bought or sold in the international exchange market are delivered on the day of transaction, a "spot transaction" is said to have occurred. The purchases and sales have taken place in the spot market. Spot transactions account for approximately 60 per cent of banks' foreign exchange transactions. The remainder are known as "forward transactions". On the forward market the buyer will be due to take delivery of goods, say, in six months' time. However, an exchange rate between the currencies is agreed at the time of the transaction. Traders are thus insuring against future fluctuations in exchange rates.

Forward buying and selling can involve either currencies or commodities on the international market. This can be achieved through a futures contract or an options contract. A futures contract could take the form of a trader agreeing, on 1 January, to purchase goods on 30 September at a predetermined price in order to avoid potentially higher spot prices. Options are the right to purchase or sell assets at some time in the future, and usually at a predetermined price. The assets may be stocks and shares, commodities, currency or property. An option is a formal contract and a price has to be paid for the option.

Example

A British company wishes to purchase machinery from the United States but pay for it in six months' time. Indications are that the pound sterling may well depreciate against the US dollar during this period. In order to avoid an increase in cost, the company can "forward buy" the US dollars required to complete the transaction. Although it will have to pay extra over the spot exchange rate for its forward dollars, it will know exactly how much will have to be paid in pounds sterling. If the company were to wait until the payment is made and purchase US dollars on the spot market, it would be gambling against future exchange rates.

2.9 Summary

Any institution or business of reasonable size has a choice of financing its activities in more than one currency in order to minimize borrowing costs. These costs should be carefully assessed in order to adopt a proper and final choice. For instance, an investment appraisal could be undertaken to formulate an appropriate decision. To finance a firm in foreign currency, the different aspects of prices, interest and exchange rates should be considered. Of specific interest here is the Fisher Equation, which relates interest rates and exchange rates and the changes in these rates. It states that differences in interest rates on similar assets denominated in different currencies are equal to the anticipated exchange rate change. However, research in this area has shown that differences between anticipated and effective exchange rates have been substantial, but essentially random. In addition, the interest rate parity theorem has tried to explain the difference in interest rates on similar assets and the difference between spot and forward exchange rates (Fisher, 1980). Further research has shown that forecasts on the forward exchange rate have not always been satisfactory. It is suggested that new decision rules are needed because of the deficiencies in current theories and methods. The research undertaken by Burmeister (1980) identified three rules that should be followed if effective borrowing is to be achieved:

- make your own forecasts in interest and exchange rate movements and select for the cheapest currency;
- arrange loans in a mixture of currencies in order to spread the risk; and
- take any currency and utilize forward cover at all times.

References

Bird, G. 1989. *Commercial bank lending and Third World debt* (London, Macmillan).

BIS Economic Papers, various years, 1985-92 (Basel, Bank for International Settlements, Monetary and Economic Department).

Boreham, G. F. 1988. "China's financial reforms", in *Canadian Banker* (Toronto), Vol. 95, No. 1, Jan./Feb., pp. 32-39.

Born, K. E. 1983. *International banking* (Leamington Spa, Warwickshire, Berg Publishers).

Burmeister, E, 1980. *Capital theory and dynamics* (Cambridge, Cambridge University Press).

Carstairs, E; Loebelson, A. 1985. "Finance giants I: The top 100 commercial banks of 1985", in *Corporate Design and Reality* (Des Plains, Illinois), Vol. 4, No. 6, July/Aug., pp. 49-55.

Collin, P. H.; Hodgson, P. 1991. *Dictionary of banking and finance* (Middlesex, Peter Collin).

Davies, A; Kearns, M. 1989. *Banking operations: UK lending and international business* (London, Pitman).

Donaldson, J. A.; Donaldson, T. H. 1982. *The medium term loan markets* (London, Macmillan).

Donaldson, T. H. 1988. *Lending in international commercial banking* (London, Macmillan).

"Eurotunnel: The financing challenge". 1987. *Euromoney* (London), Eurotunnel Supplement, June, pp. 13-14.

Fisher, D. 1980. *Money, banking and monetary policy* (Homewood, Illinois, Richard D. Irwin).

Hanson, D. G. 1985. *Dictionary of banking and finance* (London, Pitman).

Kelly, J. E. 1987. *Practice of banking,* Part 1 (London, Pitman).

Khoury, S. J.; Ghosh, A. 1988. *Recent developments in international banking and finance* (Lexington, Massachusetts, Lexington Books).

Lee, M. R. 1988. "China's Bank of Industry and Commerce takes a cautious approach to mortgages", in *East Asian Executive Reports* (Washington, DC), Vol. 10, No. 3, 15 Mar., p. 13.

Lewis, M. K.; Davis, K. T. 1987. *Domestic and international banking* (Oxford, Philip Allan).

Marsh, J. R.; Wild, D. G. 1988. *Practice of banking,* Part 2 (London, Pitman).

Perry, F. E.; Klein, G. 1992. *Dictionary of banking* (London, Pitman).

Piser, D. 1987. "Managing the construction loan", in *Bankers Magazine* (Boston, Massachusetts), Vol. 170, No. 2, Mar./Apr., pp. 51-54.

Reed, E. W. et al., 1984. *Commercial banking* (Englewood Cliffs, New Jersey, Prentice-Hall).

Robinson, J. S. 1972. *Multinational banking* (Leiden, A. W. Sijthoff).

Rosenberg, J. M. 1993. *Dictionary of banking* (New York, Wiley).

Roussakis, E. N. 1981. *International lending by US commercial banks* (New York, Praeger Publishers).

Sapp, R. W.; Watne, D. A. 1987. "Strategical and tactical issues facing US banks entering mainland China", in *World of Banking* (Rolling Meadows, Illinois), Vol. 6, No. 6, Nov./Dec., pp. 20-25.

Steinman, D. W. 1987. "Construction lending: Today and tomorrow", in *Mortgage Banking* (Washington, DC), Vol. 47, No. 10, July, pp. 89-96.

Stine, S. F. 1990. "China investment: A second wave in the offering", in *Asian Finance* (Hong Kong), Vol. 18, No. 6, 15 June, pp. 13-14.

Walker, I. S. 1985. "Protecting against risk when closing a construction loan", in *Journal of Commercial Bank Lending* (Philadelphia, Pennsylvania), Vol. 68, No. 2, Oct., pp. 15-23.

Weston, R. 1981. *Domestic and multinational banking* (London, Croom Helm).

World Bank 1991. *Glossary of finance and debt* (Washington, DC).

SYNDICATED COMMERCIAL BANK FINANCING

3

3.1 Introduction to syndicated loans

Syndicated loans are the most important risk-sharing device suitable for large loans. They involve the formal subcontracting of the loan and the associated risk, and are considered to be the same as co-insurance and re-insurance arrangements. Thus, individual banks (or institutions) can spread the risks associated with large individual projects and avoid overexposure. Syndicated loans have benefits for both the client and the banks. They enable loans to be made for single projects whose funding requirements make them too large for single banks. Because the overall risk is reduced, so is the overall cost of the finance.

The borrower negotiates the loan conditions with one of the banks which acts as the "lead manager". It is the responsibility of the lead manager to locate other banks which are prepared to offer (on common terms) a medium-term, syndicated, rollover (credit available over a continuing period) Eurocurrency loan to a single borrower. There are many different forms of syndicated loan, which are grouped by the reason for the loan, type of syndication, drawdown method (drawing money which is available under a credit agreement), and repayment terms, but they are all linked by the following key elements:

- **duration**. The duration of this type of loan is usually three to 15 years, i.e. a medium-term loan.
- **rates of interest**. Interest rates are usually quoted as a percentage above the London Interbank Offered Rate (LIBOR) or the US Prime Rate, which will be adjusted at pre-set intervals of, say, six or 12 months.
- **currency**. The most common currency used is the Eurodollar, or in some cases a "basket" of currencies. In addition, the borrower may be given the choice of changing to other currencies at rollover.

The syndicated loans made world-wide in 1987 and 1991 are presented by type of currency in table 3.1.

Table 3.1. Syndicated loans by type of currency, 1987 and 1991 (percentages)

Currency	1987	1991
US dollar	59.4	85.5
ECU	2.9	4.2
Pound sterling	18.0	3.2
Deutschmark	2.9	1.9
Swiss franc	0.8	0.6
Japanese yen	11.8	0.1
Others	4.2	4.5
Total	100.0	100.0

3.2 Types of syndicated loan

- *Traditional syndicated bank loans (floating rate)*

  Traditional syndicated bank loans are based on a variable rate but with a fixed maturity, drawn once, and repayments made to an agreed schedule. A period of grace is often included. The loan will usually be negotiated by a single lead bank which will form a syndicate with other banks. The loan will have to be administered and this will often be done by the lead bank.

- *Syndicated bank loan (fixed rate)*

  Syndicated fixed-rate bank loans are very similar to traditional loans, but the interest rate remains fixed throughout the term of the loan.

- *Revolving credit*

  Revolving credits have similar characteristics to syndicated bank loans but permit the borrower to draw the loan (or part of it) and make repayments at his or her discretion, or to an agreed programme, throughout the term of the loan.

- *Standby facility*

  Traditional syndicated Eurocredits usually require the borrower to draw down the loan within a set time period. A standby facility allows the borrower to delay taking the loan until he or she wishes to do so. A contingency fee has to be paid until the loan is drawn and interest starts to be charged.

- *Multi-option facilities (MOF)*

  Syndicated loans with MOF are simple to operate and cheap to obtain. The facilities that are provided tend to be much more complex than traditional syndicated loans, but can be tailored to suit the individual borrower's requirements. For example, a MOF could include a medium-term standby loan which would guarantee future funds if the market dried up, while cheap funds could be obtained through the short-term money markets.

3.3 Parties to a syndicated loan

The parties to a syndicated loan agreement can be placed into the following categories, depending upon the role they perform:

- The **borrower** must provide the lead manager with full financial statements throughout the loan period. Prior to drawdown, the borrower must agree to certain conditions: for example, an assurance that there is no default on existing loans and a guarantee that the nature of business will not change.
- The **lead manager** is the bank responsible for negotiating the terms of the loan and, if appointed as the agent bank, will administer the loan.
- The **managers,** along with the loan manager, organize the syndicated participants, and fully or partially underwrite the loan.
- The **participants** are usually smaller banks which the lead manager approaches to join the loan. They may not have the experience to act as a manager, but share the loan and thus the non-underwriting risk.
- In the United States and Euromarkets, the function of the **agent** is usually delegated to another bank. The agent performs the functions of gathering funds from the syndicate members once notification of drawdown has been made, collecting interest and repayment from the borrower, and subsequent distribution to the syndicate members.

3.4 Steps in completing a syndicated loan

The process of arranging a syndicated loan must be based on well-founded procedures. It is essential that each loan is tailored to satisfy the borrower's needs, and as a result every deal is different. In the early stages of the deal, there must be considerable discussion between the client and the various sections of the lead bank, through the borrower's account manager. The type and number of questions raised will vary between borrowers, but the following issues will have to be resolved:

- the number of banks to be included in the underwriting group;
- the proportion of the lenders to be relationship and non-relationship banks (the former deal directly with the borrower and handle the provision of services, while the latter are effectively the subcontractors);
- the loan conditions; and
- when the deal should be launched.

During this stage of the negotiation, the borrower should receive clear instructions from the lead bank as to the actions he or she should and should not be taking. A good syndication strategy (as agreed by the lead bank and the borrower) is essential if the borrower's needs are to be satisfied through a successful market reception. Although each deal has to be adjusted to the borrower's needs, there are four main steps in completing a syndicated loan. These are listed below:

- discussions between the borrower and the lead bank in order to establish the terms of the deal;
- production of an information memorandum and preparation of the loan syndication by the lead bank;
- negotiation of the loan agreement; and
- finalization of the deal.

The various steps in completing a syndicated loan are illustrated in figure 3.1.

Figure 3.1. Steps in completing a syndicated loan

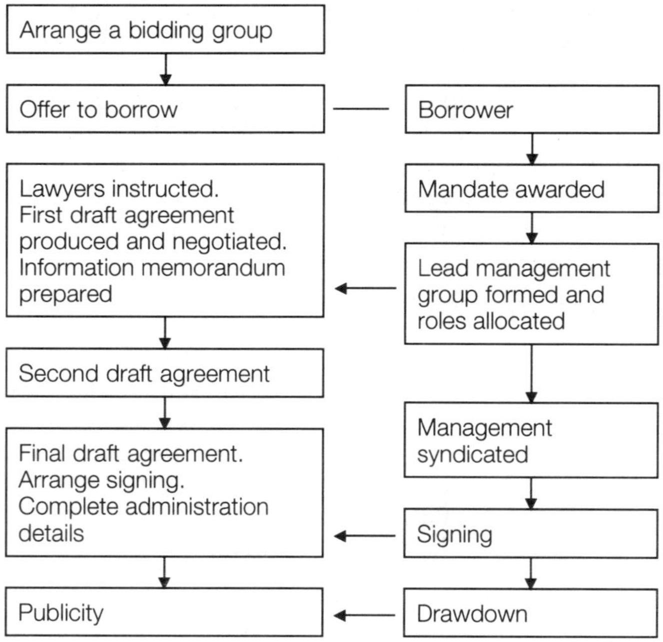

3.5 Loan documentation

The various terms of the loan and the respective roles and duties of the parties are set out in the "loan agreement", which forms the contractual basis of the advance. As the experience of both lenders and borrowers has grown over the years, so has the complexity of the legal documents surrounding the loan. In the early 1970s, most loan agreements were only six or seven pages long; however, current practice often requires between 40 and 90 pages of documentation. A typical loan agreement usually contains the following sections:

- the parties and their obligations;
- definitions;
- default;
- fees and assignment;
- right of set-off (to use debt owed to one party to reduce a debt owed to them);
- the agent's role;
- governing law;
- Schedule I: Names and addresses of participating banks and the extent of their participation;
- Schedule II: Legal authority of the guarantor to undertake this liability;
- Schedules III-V: Opinions of the legal advisers to the borrower, guarantor and agent concerning the validity of the loan agreement and the guarantee;
- choice of governing law; and
- choice of court's jurisdiction.

In order to avoid later uncertainty, the loan agreement must express the choice of law, and the jurisdiction of certain courts should be set out. Very often, the governing law of the loan is English.

The lead manager has to fix an acceptable rate for the loan. This must take into account the responsibility to the borrower to ensure that the loan is competitive when compared to other sources of finance, as well as attractive to the lenders. The cost of a loan comprises:

- the fees; and
- the spread or margin (interest charged).

The lead manager receives a management fee for arranging the loan (this figure is not usually disclosed). The agent is paid an annual fee for administering the loan, which may be expressed as a single sum or as a percentage. Participating banks will also charge a fee to cover any part of the loan which is not taken up. The spread (or margin) is usually based on the interbank rate for the currency of the loan and is quoted as a figure over LIBOR, US Prime Rate or SIBOR (Singapore Interbank Offered Rate).

3.6 Background to Euromarkets

In the 1950s, there was an increase in the number of non United States residents who held considerable sums of US dollars, and the Bank of England was restricted in the granting of sterling credits. The restrictions placed on the British banks encouraged them to make US dollar advances for the financing of international trade. Currency regulations in the early 1960s made the rates for Eurodollar deposits more favourable. In July 1963, the Interest Equalization Tax effectively increased the cost of foreign bonds issued in New York. It thus became advantageous to issue foreign borrowers with dollar bonds through European Financial Centres which held deposits of US dollars.

In 1933, Regulation Q was introduced in order to reduce the cost of housing finance in the United States. This was achieved by placing a ceiling on the rate of interest which American commercial banks could pay on term deposits. Commercial banks could thus attract investors even without offering low interest rates and, consequently, banks could provide loans at very reasonable rates. This measure encouraged the growth of the Eurodollar market and an outflow of funds from the United States. In 1980, it was decided to phase out Regulation Q over a five-year period and this, along with a reserve imposition on American banks in respect to foreign lending, resulted in a large flow of funds back to the United States. On the other hand, the oil-exporting countries provided a good supply of Eurodollar deposits which enabled the increased demand from borrowers to be satisfied.

Most Eurocurrency lending in the 1960s was in the form of Eurobonds issued by companies, but this demand fell in the 1970s along with an increase in borrowing requirements by governments to fund balance-of-payment deficits or finance large projects. The size of these loans was so great that banks sought to spread the risk by the process of syndicated lending. In particular, the developing countries with no oil to export found syndicated loans a valuable source of finance when dealing with balance-of-payment deficits.

The increase in demand for syndicated loans resulted in a large number of banks appearing in London, encouraged by the relatively small number of monetary restrictions made upon the Euroloan market. This resulted in increased competition, and in 1973 margins and fees were reduced to very unattractive levels.

Bankhaus I.D. Herstatt, a major player in the Eurodollar market, collapsed in May 1974 and confidence in the interbank market dropped to such an extent that depositors moved funds from the Eurodollar market to the United States. As a result, short-term interest rates rose dramatically. The outcome was that the Eurodollar market became tiered, with the larger international banks being able to obtain funds at a lower rate than the smaller banks. This enabled confidence to be restored, and only a year later interest levels fell to their previous levels. As a result, more types of borrower entered the market and a large number of existing borrowers were able to renegotiate loans which were taken out at less favourable rates.

The Euromarket received its second major setback in 1979, when events in Afghanistan and the Islamic Republic of Iran made it necessary for the American monetary authorities to support the dollar, which led to a sudden rise in short-term interest rates. Non-prime borrowers were given higher spreads and new borrowers were given shorter maturities. In addition, non-oil-producing developing countries, particularly those which relied heavily on the export of commodities, were badly hit by the world recession and depressed prices of raw materials. In many cases it became difficult to service large commercial loans, and many borrowers were forced to reschedule their repayments. This caused a drop of 25 per cent over two years in the number of new loans being issued.

In the mid-1980s, there was a definite move by corporate borrowers towards note issuance and revolving underwriting facilities (NIFs) with very attractive pricing terms. (NIFs are promises by banks to lend money to companies when it cannot be raised on the short-term securities markets; the revolving underwriting facility is an extension of the NIF and takes the form of a conventional bank loan at low short-term interest rates offered when the money market fails to purchase all the short-term commercial paper offered.) This resulted in several banks lowering or even closing their loan syndication units. During 1987 the Eurobond market shrank and lost the ground it had made during 1986. However, new profit areas were provided by an increase in the Euronote and syndicated loan markets. The introduction of multiple-option facilities between 1986 and 1988 caused banks to redirect their interests and create specialist teams of loan syndicators. There was also a significant move away from banks being appointed as joint lead managers of facilities, and many more sole lead mandates are now being awarded.

The indications are that the syndicated loan market will shrink during the 1990s. For example, Manufacturers Hanover topped Euromoney's ranking with US$23.3 billion worth of loans arranged during the first six months of 1989. However, for the same period in 1990 this dropped to only US$4 billion.

3.7 Recent trends

The top 25 syndicated loan providers and top 25 loan arrangers for the first six months in 1990 are presented in tables 3.2 and 3.3. When these are compared with the figures for the first three months of 1988 presented in tables 3.4 and 3.5, the dramatic fall in the size of the syndicated loan market becomes apparent.

The recent problems of the syndicated loan market have resulted in some innovative approaches being taken by project owners in order to secure the funds they require. The case studies presented in this chapter demonstrate recent success stories for syndicated loan arrangements stemming from very creative deals.

Table 3.2. Top 25 syndicated loan providers, 1 January - 30 June 1990

Provider	Amount (US $ million)	No. of loans	Share (%)
NatWest	5 191.41	195	2.97
Fuji	3 621.57	136	2.08
Barclays	3 562.20	150	2.04
Citicorp	3 362.59	88	1.93
Sumitomo	3 212.64	130	1.84
Bank of America	2 805.96	98	1.61
Chemical	2 773.32	56	1.59
Security Pacific	2 578.08	61	1.48
Chase Manhattan	2 395.26	99	1.37
Crédit Suisse	2 393.28	68	1.37
Industrial Bank of Japan	2 355.05	82	1.35
Mitsubishi Bank	2 283.19	118	1.31
First National Bank of Chicago	2 276.87	78	1.30
Crédit Lyonnais	2 212.34	112	1.27
J.P. Morgan	2 196.56	61	1.26
Swiss Bank Corp.	2 088.86	66	1.20
Mitsui Taiyo Kobe	2 064.92	99	1.18
Bankers Trust	1 991.64	57	1.14
Union Bank of Switzerland	1 970.13	80	1.13
Sanwa	1 960.23	105	1.12
Bank of Tokyo	1 932.51	97	1.11
Manufacturers Hanover	1 896.83	85	1.09
Dai-Ichi Kangyo	1 769.31	92	1.01
Commerzbank	1 651.35	69	0.97
Bank of Nova Scotia	1 659.00	68	0.95

Source: *Euromoney*, Aug. 1990, p. 32.

Table 3.3. Top 25 syndicated loan arrangers, 1 January - 30 June 1990

Arranger	Amount (US $ million)	No. of loans	Share (%)
Citicorp	16 236.70	54	9.30
J.P. Morgan	13 205.30	27	7.57
Bank of America	12 239.70	40	7.01
Chemical	9 557.12	20	5.48
First National Bank of Chicago	8 187.47	45	4.69
Chase Manhattan	7 479.29	62	4.29
Barclays	6 977.12	41	4.00
S.G. Warburg	5 483.06	15	3.14
Chevron	5 000.00	1	2.86
Bankers Trust	4 867.55	40	2.79
NatWest	4 223.16	39	2.42
Manufacturers Hanover	3 965.11	48	2.27
Security Pacific	3 184.39	25	1.82
Continental	3 026.22	20	1.73
Crédit Suisse	2 185.50	11	1.25
Bank of Tokyo	2 156.19	11	1.24
Mitsubishi Bank	1 979.52	25	1.13
Svenska Handelsbanken	1 881.85	2	1.08
Deutsche Bank	1 869.71	6	1.07
Swiss Bank Corp.	1 765.88	12	1.01
Lloyds Bank Capital Markets	1 711.12	14	0.98
Westpac	1 710.14	11	0.98
Toronto-Dominion	1 688.35	6	0.97
Hong Kong Bank Group	1 613.94	14	0.92
Sumitomo	1 583.25	17	0.91

Source: *Euromoney*, Aug. 1990, p. 32.

3.8 Case study 1 : Pechiney

In 1990, Chase Investment Bank had the task of arranging finance for an aluminium smelting plant to be operated by Pechiney in France. The state-owned Electricité de France (EDF) had agreed to provide cheap electricity to the plant for 25 years, in return for a proportion of the profits. Pechiney was to own 35 per cent of the company which was to be set up for this purpose, with most of the remaining equity being placed among French institutions. The total cost of the project was US$1.1 billion, with Pechiney and EDF providing subordinated advances of US$420 million. Thus, Chase Investment Bank had to secure loans worth US$680 million. A group of banks already associated with Pechiney was asked to provide US$30 million. By the time the response period was up, Chase Investment Bank had received offers totalling US$2.7 billion from 43 banks in the United States, Japan and Europe.

The resulting loan was over a 15-year period with 6.25 base points over LIBOR. Higher margins up to 7.5 base points were allowed for,

Table 3.4. Top 25 syndicated loan providers, 1 January - 31 March 1988

Provider	Amount (US $ million)	No. of loans	Share (%)
Banque National de Paris	2 567.20	72	2.69
Barclays de Zoete Wedd	2 404.99	84	2.52
Crédit Suisse	2 052.24	49	2.15
Deutsche Bank	1 658.42	31	1.74
Crédit Lyonnais	1 643.05	81	1.72
Swiss Bank Corp.	1 628.64	28	1.70
Industrial Bank of Japan	1 616.08	55	1.69
NatWest	1 461.20	64	1.53
Société Générale	1 381.64	60	1.45
Commerzbank	1 334.00	46	1.40
Fuji Bank	1 330.78	49	1.39
Midland Montagu	1 296.88	44	1.36
Bank of Tokyo	1 287.07	45	1.35
Sanwa Bank	1 284.91	70	1.34
J.P. Morgan	1 215.35	19	1.27
Mitsubishi Bank	1 193.02	51	1.25
CIBC	1 125.65	37	1.18
Sumitomo Bank	1 108.74	46	1.16
Union Bank of Switzerland	1 103.15	10	1.15
Amsterdam-Rotterdam Bank	1 097.93	33	1.15
Citicorp	1 084.71	32	1.13
Chase Investment Bank	1 084.29	41	1.13
Bank of Nova Scotia	1 083.44	31	1.13
Dresdner Bank	1 078.71	29	1.13
Royal Bank of Canada	979.47	32	1.02

Source: *Euromoney*, May 1988, pp. 1-3.

depending upon future cash flows. This was achieved by producing a structure which kept borrowing off Pechiney's balance sheet, while at the same time persuading lenders that Pechiney was backing the project.

3.9 Case study 2 : The Channel Tunnel

The Channel Tunnel linking France and the United Kingdom is one of the largest infrastructure projects ever undertaken. There were many early doubts as to the viability of financing a project of such a magnitude via private sources. However, a consortium of 198 international banks, led by five lead banks, committed over US$5,000 million to the project. The deal was signed in November 1987, and was for an 18-year loan period with standby facilities. Further details relating to the Channel Tunnel are presented in Chapter 9.

3.10 Case study 3 : Markets in Malaysia

The Malaysian Government is determined to push forward the country's development plans and create a fully developed State by the year 2020. Planned development expenditure for the next ten years exceeds

Table 3.5 Top 25 syndicated loan arrangers, 1 January - 31 March 1988

Arranger	Amount (US $ million)	No. of loans	Share (%)
J.P. Morgan	12 751.50	22	14.42
Citicorp	7 862.73	20	10.02
Bankers Trust Company	6 548.20	12	7.41
Swiss Bank Corp.	6 355.00	5	7.19
Manufacturers Hanover	5 220.38	17	5.90
Security Pacific Bank	3 868.88	10	4.38
Barclays de Zoete Wedd	3 708.63	17	4.19
Chemical Bank	3 381.90	7	3.82
Chase Investment Bank	2 947.84	26	3.33
NatWest	2 663.08	19	3.01
Bank of America	2 509.99	12	2.84
Crédit Suisse First Boston	2 344.51	7	2.65
Kleinwort Benson	1 800.55	4	2.04
S.G. Warburg & Co	1 287.19	11	1.46
Midland Montagu	1 212.45	9	1.37
Crédit Lyonnais	1 119.79	13	1.27
Hong Kong Bank Group	1 092.25	4	1.24
Banque Nationale de Paris	1 070.38	13	1.21
ANZ-Grindlays Banking Group	961.75	4	1.09
Société Générale	941.64	14	1.06
First National Bank of Chicago	901.82	4	1.02
Merrill Lynch International Bank	897.50	5	1.01
Morgan Grenfell	873.28	7	0.99
Bank of Tokyo	860.24	6	0.97
J. Henry Schroder Wagg	679.40	3	0.77

Source: *Euromoney*, May 1988, pp. 1-3.

US$90 billion. There are currently US$14 billion worth of construction projects under way, with plans for another 32 projects valued at US$24 billion. Government policy has been to direct most infrastructure projects towards the private sector, which has created a high demand for large-scale commercial credit. In the past, foreign bankers have had to compete with a relatively strong domestic banking sector for the limited number of suitable projects, and consequently margins have been small. However, the traditional good risk-to-reward ratio when compared with the rest of South-East Asia, and the increased activity could well prove very attractive to foreign banks.

The high cost of the planned development requires that a large proportion of the projects will have been financed by limited recourse funding. This usually involves loans being made to organizations which do not have sufficient assets to guarantee repayment of the loan should problems occur; consequently the bank takes more of the risk involved. It is inevitable that only well-established lenders will have the experience to deal with limited recourse projects on such a scale. Obviously, the high demand for funds will push up the relatively low margins. However, the good risk-to-reward ratio should restrict margins to a reasonable level.

Syndicated lending to Malaysia has been high over recent years. Prior to 1992, over US$10.8 billion worth of syndicated loans had been provided to Malaysia from foreign lenders. About a quarter of these went to the construction sector. Most have been medium-sized, structured loans for large government corporations with government backing. However, project promoters are finding it difficult to accept the increase in pricing that comes with the shift to limited recourse funding.

Wherever possible, the Malaysian Government has encouraged onshore participation in most of the loan syndications, particularly in an effort to reduce foreign borrowing, but also to improve the expertise of the Malaysian banking sector. For example, the deal to finance the second and third phases of a petrochemical plant in Johor comprised the following elements:

- *Offshore tranche*: US$166 million charged at 1.25 per cent above LIBOR on a ten-year maturity.
- *Onshore tranche*: US$128 million charged at 1.25 per cent (1.5 per cent after completion) above LIBOR on a 20-year maturity.

3.11 Summary

The main advantage of syndicated loans is their flexibility. The banking market has often proved to be the preferred source of funding as compared to the securities market. The reasons for this are:

- speed of execution;
- funds available;
- reduced administration; and
- flexibility of timing.

Over the years, there has been considerable speculation over the impending demise of the syndicated loans market. However, when the conditions are favourable, remarkable oversubscriptions are possible on certain types of project, for example, the aluminium smelting plant operated by Pechiney in Dunkirk, France.

Several factors have contributed to the drop in number of syndicated loans being arranged, some of which are listed below:

- The period of high-leverage buy-outs has started to draw to a close. Multibillion-dollar loans are no longer required for American take-overs (i.e. loan buy-out).
- Real estate related lending in the United States and the United Kingdom has been reduced as a result of falling property prices.
- Banks must now meet the guidelines on capital adequacy of the Bank for International Settlements and are forced to ensure that they receive

reasonable returns from loans which have to be supported by capital.

- Japanese banks have been hit by the decline in the Tokyo Stock Market, and their traditional support for syndicated loans has been substantially dampened.
- The current world recession.

References

Bennett, R. 1993. "Project finance: Morton's folly", in *Euromoney,* June, pp. 59-60.
Euromoney (London). 1992. "Syndicated loans: Queuing up to lend", Malaysia Supplement, Aug., pp. 71-77.

Hagger, E. 1992. "Arriverderci lender", in *Euromoney*, Aug., pp. 63-67.

Howcroft, B. 1985. *Syndicated lending by banks* (Cardiff, University of Wales Press).

Humphreys, G. 1988a. "Syndicated loans: Cut-throat game of pass the parcel", in *Euromoney*, Syndicated Loans Supplement, May, pp. 20-24.

——. 1988b. "Syndicated loans: Volume goes up – but profits?", in *Euromoney*, Syndicated Loans Supplement, May, pp. 10-16.

Jones, R.; Barrett, M. 1988. "The best of times – The worst of times", in *Euromoney*, Markets 1988 Supplement, Mar., pp. 1-48.

Keller, P. 1989. "Lend the money, then sell the debt", in *Euromoney*, Aug., pp. 99-106.

Kochan, N. 1989. "Lenders step into the breach", in *Corporate Finance Supplement*, Dec., pp. 23-25.

Lee, P. 1990. "Syndicators feel the chill", in *Euromoney*, Aug., pp. 31-34.

Pavey, N. 1988. "Syndicated loans: Razor-thin returns on cheap standbys", in *Euromoney*, Syndicated Loans Supplement, May, pp. 1-3.

Schares, G. E. 1989. "The East bloc's $100 billion phone bill", in *Business Week* (Industrial/Technology Edition), No. 3134, 20 Nov., pp. 139, 142.

PART II

SOURCES OF INTERNATIONAL PROJECT FINANCE

PROJECT FINANCING FROM COMMERCIAL SOURCES 4

4.1 The role of the sponsor

Project sponsors may include government departments, international agencies, public corporations or parastatal entities. The sponsor's objectives may include political, material inputs, marketing, technical, management and financial aspects, but they usually relate to the sponsor's wish to exploit its particular resources. The owner of the project and the project sponsor do not have to be the same, and on many major capital projects the shareholders of the sponsor can only guarantee the large financial resources required by using the project's future cash flow, hence the term "project finance". The aim of most project financing is to arrange borrowing for a project which will benefit the sponsor and at the same time be completely non-recourse to the sponsor (i.e. in no way affect its credit standing or balance sheet). Hence, project financing is often referred to as off-balance-sheet financing. Few projects are financed independently on their own merit without credit support from sponsors or other interested third parties.

This chapter initially presents the types of capital generally available to finance projects and goes on to show how these can be used as off-balance-sheet or on-balance-sheet loans. The importance of guarantees in securing project finance is demonstrated. The chapter concludes with a case study of Terminal 2 at Manchester Airport in the United Kingdom.

4.2 Types of capital

There are three main types of capital available to finance projects through either off-balance or on-balance-sheet finance. These are explained below.

"Equity" investment represents the risk capital and most junior debt. It provides the basis for lenders or investors to advance more senior forms of capital to the project.[1]

Quasi-equity represents the loans or advances made to a project, and are senior to equity capital, but junior to senior debt and secured debt.

Senior debt can take the form of either secured or unsecured loans. Most borrowings from commercial lenders for project financing will be in the form of senior debt. Senior debt has priority of payment if the borrower gets into financial difficulty. Commercial loans are the most important source of senior debt for project financing and may involve a single lender, several lenders or a loan syndicate.

These types of capital may be in the form of construction loans, long-term loans, or working capital loans. The following details should be covered in the loan agreement:

- the amount which may be borrowed;
- commitment fees for overdraft or term loans that are not fully used;
- term of the loan and repayment schedule;
- interest rate on the outstanding balance;
- procedure for take-downs and conditions precedent for the take-down;
- representations and warranties (i.e. guarantees that loans will be available) of the borrowing;
- legal opinions which will be required at the closing of the loan agreement;
- affirmative covenants (i.e. agreements relating to specific actions to be taken);
- financial covenants (i.e. financial agreements);
- responsibility for any withholding of tax on interest;
- enforceability of the rights of the lender.

The element of risk associated with the provision of a loan must be reflected in the interest rate charged. Moreover, there are several mechanisms available that can be used to reduce the cost of borrowing by decreasing the lender's risk.

4.2.1 Unsecured loans

Unsecured loans are available when the debt is backed by the general credit of the borrower, and is not secured by a security interest in any asset or pool of assets. This is often referred to as "name lending". Unsecured loans are available to projects whose sponsors, owners and managers have established good reputations with the financial community and where sufficient capital or subordinated loans have been provided to meet the equity risk capital requirements of the project.

4.2.2 Secured loans

Secured loans are available to most projects where the assets securing the debt have value as collateral; such assets must be marketable and readily convertible into cash. The collateral of real property, personal property, payments due under a take-or-pay contract (see subsection 4.8.1) and assignment of contractual rights are all used as collateral under project financing.

4.2.3 Syndicated loans

Commercial banks are the largest source of project loans, but tend to limit their commitment to between six and ten years, with floating interest rates based on LIBOR or the US Prime Rate. However, longer loan periods can be negotiated and fixed interest rate loans for five- to ten-year maturity are sometimes available.

The size of some projects is such that the large loans required to finance them have to be provided through syndicated international commercial banks (see Chapter 3). The general advantages of the syndicated loan market are that:

- substantial amounts of debt can be raised;
- loans may be made in any of several currencies;
- risks can be shared;
- banks involved in syndicated loans are sophisticated and are able to understand and participate in the complex credit risks presented by project financing;
- the take-up date of loan can be flexible; and
- pre-payment is customarily permitted.

4.2.4 Commercial finance companies

Large commercial finance companies are another potential source of funds for project financing. Compared to banks or insurance companies, finance companies do not have a depositor base of policy-holders as a source of funds. They must buy all their funds in the debt markets and re-lend at a spread. Consequently, funds from commercial finance companies tend to be highly priced and limited in volume. Other sources of fund-raising include leasing companies, and so on.

4.3 Normal balance sheet financing

Banks providing normal balance sheet lending must satisfy themselves that the borrower's total financial structure is sufficient to pay interest as well as repay the loan. In normal balance sheet financing, a company will borrow money from a bank, the debt will appear on the company's balance sheet, and in most cases the assets shown on the company's balance sheet will act as the bank's security. However, if the company wishes to invest in a specific project, it is possible to safeguard the company's assets by creating an off-balance-sheet arrangement; this is explained in section 4.4.

Some of the instruments used to raise funds to finance projects are listed below:

- commercial sources (equities and debts: commercial banks, etc.);
- syndicated bank and Eurodollar loans;
- Eurobonds;
- Eurodollar bonds;
- leases;
- commercial paper market.

4.4 Pure project financing

There are many different financing options available, but lenders will usually require some type of security to be offered. Pure project financing will only occur when the loan's security is based totally on the project's cash flow. Owing to the complex nature of most projects, it is rarely possible to achieve this situation. Lenders providing project financing must therefore ensure that there will be enough cash flow generated by the project to pay interest and repay capital after all other expenses have been met (e.g. running costs, taxes). In project financing, the lender will therefore evaluate the specific project in detail, rather than the activities of the sponsor. Project financing is sometimes referred to as off-balance-sheet finance or non-recourse finance.

The reasons for selecting project financing to achieve the sponsor's objectives could be to:

- enable the sponsor to undertake a project that would have otherwise been impossible and to receive the benefits of a large capital asset without having to finance it;
- minimize the sponsor's equity contribution by attracting other investors from either the private or the public sector;
- shift some of the risk away from the sponsor on to the project;
- help the sponsor to raise loans on better terms than would have been achieved through direct borrowing; and
- keep the project loan off the sponsor's balance sheets and achieve a higher level of gearing.

The use of these terms can sometimes be misleading. "Off-balance-sheet finance" means that the finance is not provided by a loan secured against the balance sheet assets. "Non-recourse finance" means that there are no loan repayment guarantees provided by either the borrower or the parent company.

The use of guarantees can often give lenders a false sense of security. It should never be assumed that a guarantee can easily be enforced. Guarantors can find many excuses for non-payment and the lender must reserve its rights. The recent failures in Italy, where the Government is refusing to accept liability for the still outstanding loans of one of its wholly owned subsidiaries, highlight the problem.

4.5 Performance bonds

With any major project come the inherent risk factors associated with accidents on site, additional construction costs, completion delays and future cash flows. It is vital that bankers or sponsors reduce these risks or share them out with others through the correct guarantees, insurances and bonds. Several types of bond are available to ensure completion or performance under a construction contract, and these can be used to provide some extra assurance to the project lender. The most common bonds are discussed in more detail, along with other forms of insurance, in Chapter 10, while the rest of this chapter discusses guarantees of a more commercial nature.

4.6 Owner/sponsor guarantees

The owner of a project is the most obvious guarantor of a project financing transaction. However, subsidiaries established to run a project are often undercapitalized and have no established track record, leading to poor credit rating. Project lenders will therefore require guarantees from a creditworthy source, often involving the parent company. In most cases, this type of debt guarantee will appear as a liability on the consolidated balance sheet of the owner or sponsor. However, there are other forms of direct and indirect undertaking which, if structured correctly, could be treated by the sponsor as off-balance-sheet liabilities.

Sponsors may or may not be interested in owning a facility which is needed to provide it with a product or service. Figure 4.1 illustrates the process which will result in the sponsor owning the completed facility. Figure 4.2 illustrates the mechanism that the sponsor must establish if it does not wish to own the completed project but is prepared to act as a guarantor. The arrangements illustrated in figures 4.1 and 4.2 are both direct guarantees.

Figure 4.1. Sponsor guarantee for delivery

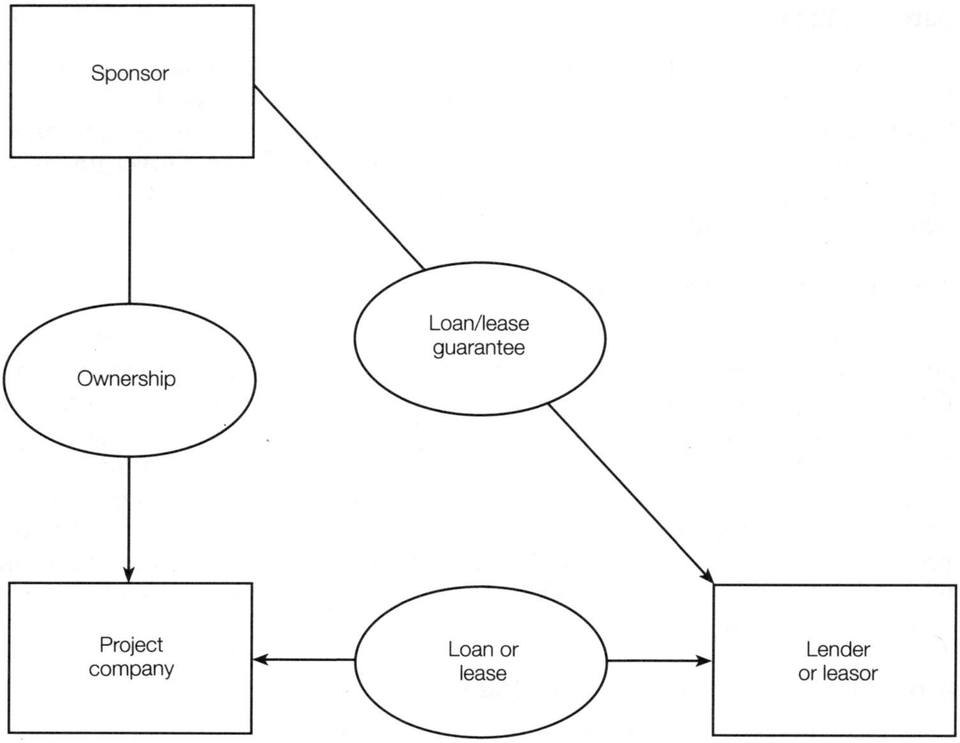

Figure 4.2. Sponsor guarantee for external company

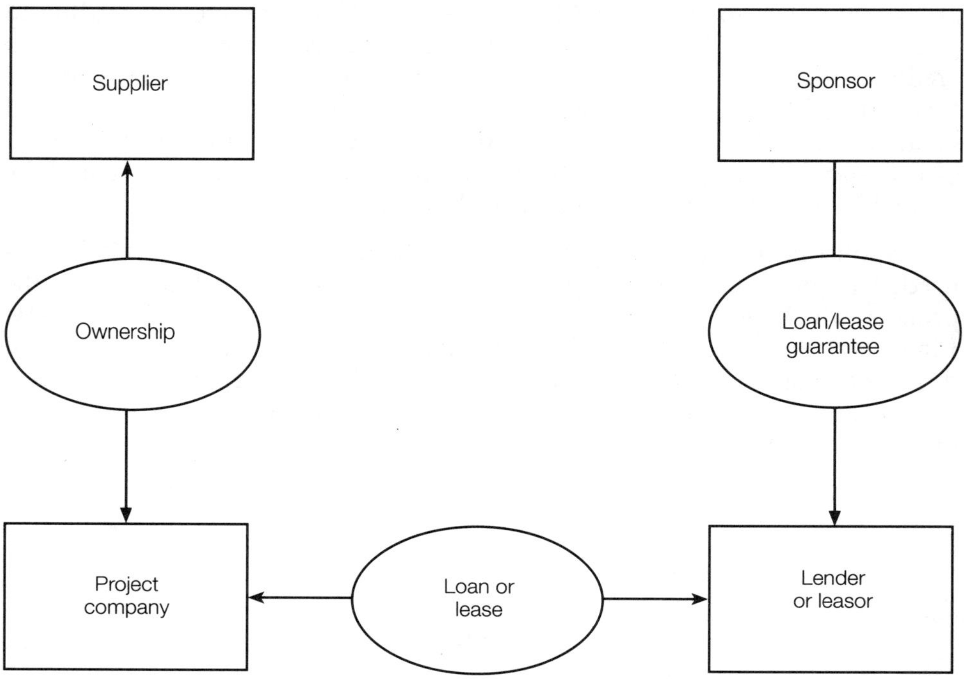

4.7 Third party direct guarantees

One objective of project financing is to ensure that no single party carries the financial burden or risk associated with the project. This can be achieved by combining different types of guarantee or undertaking from various parties in the form of a bankable credit. Guarantees enable sponsors to transfer the financial risk of a project to a third party. This mechanism allows off-balance-sheet financing to take place, without which some projects would not be possible.

Third party guarantors must benefit from the transaction in order to make their involvement worthwhile. Any guarantor is thus, to some extent, a project sponsor. The advantage of third party guarantees is that the owner or sponsors (non-guarantor) are able to keep their liabilities off balance sheet. It is possible to categorize third party guarantors into the following groups:

- suppliers;
- sellers;
- users;
- contractors; and
- government agencies.

The loan/lease obligations of the project company are guaranteed by a third party who does not wish to own or control the project company. Based on this guarantee, the project company is able to enter into a loan or lease agreement with the lender or leasing company. The relationships involved are illustrated in figure 4.3.

Figure 4.3. Third party direct guarantee

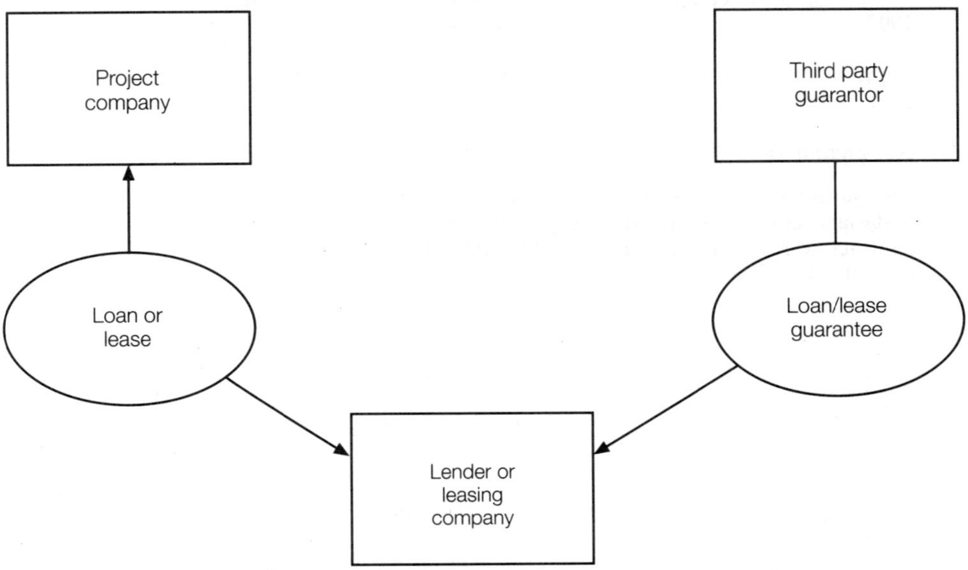

4.8 Third party indirect guarantees

Third party indirect guarantees can take the form of "take-or-pay contracts" or "throughput contracts", and can be used to ensure a steady flow of funds once a project has been completed. These may include a supplier setting a fixed price for raw material over a predefined period, or the users of the project's output agreeing to take a set quantity of goods at a fixed price. These types of arrangement are often sufficient to assure lenders of the project's viability. A trustee (usually a bank) should act as an intermediary and, by controlling the flow of money, provide additional assurance to the lender.

4.8.1 Take-or-pay contracts

Figure 4.4 illustrates a typical take-or-pay contract. Consider a concrete manufacturer who intends to increase overall production by building a new factory. In order to secure the funds required to finance this project, the manufacturer has entered into a take-or-pay contract with a subsidiary of Company A. Under this contract, the subsidiary has agreed to purchase 40 per cent of the manufacturer's planned production for the period that the debt used to build the factory remains outstanding. The payments made by the subsidiary equal 40 per cent of material costs, operating costs, depreciation, interest on capital used to build the factory and a return on owner's equity. The following disclosure, or similar, must appear in Company A's accounts:

> In order to guarantee supply on a long-term basis, it has been necessary for one of the company's subsidiaries to enter a contract and agree to purchase 40 per cent of the output of a concrete manufacturer through to the year 2015, and to make the following annual payments whether or not it is able to take delivery of the concrete.

	$ (thousands)
1992-97 ($7,000 per year)	35 000
1997-2015	133 000
Total	168 000
Less: (interest)	62 000
Net present value	106 000

> The subsidiary must also reimburse the concrete manufacturer for its share of raw materials, costs and operating expenses. The subsidiary's total purchases under this contract were (in thousands) $10,000, $11,000 and $12,000 in 1989, 1990 and 1991, respectively.

Figure 4.4. Take-or-pay contract

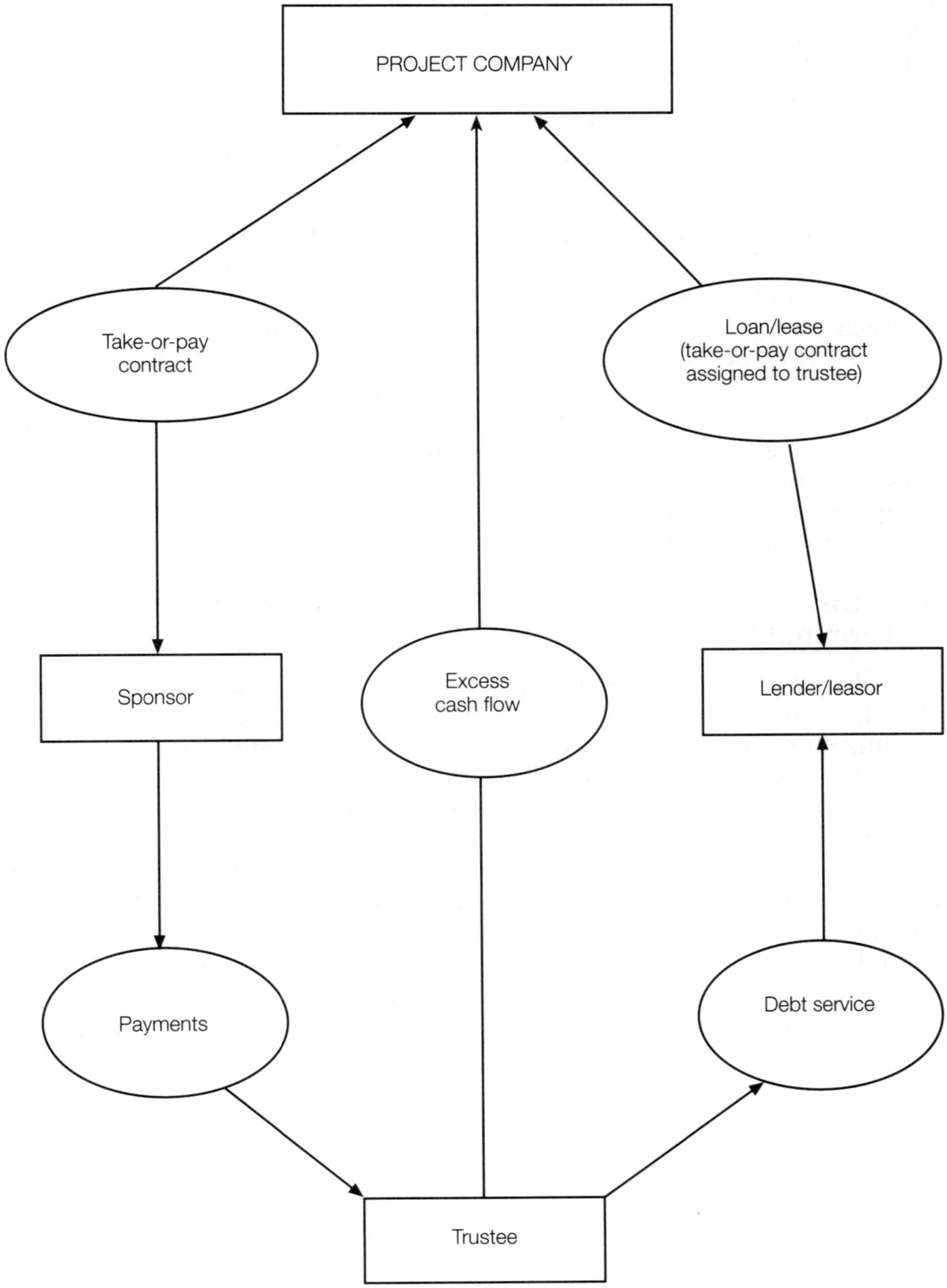

4.8.2 Throughput contracts

A throughput contract is usually an agreement between the owners of a transportation or processing facility and the potential users of that facility, (e.g. an oil pipeline), whereby the users agree periodically to pay a set amount in return for the processing or transportation of a product. The users should provide an agreed minimum quantity for each period and have to pay even if the contracted quantity is not provided. Throughput contracts signed prior to construction can act as guarantees to future income and thus help to raise the funds required to finance the project.

Figure 4.5 illustrates an arrangement for a typical throughput agreement. This agreement has been used to secure a loan which will be used to build a processing plant. A construction contract is formed between the processing company and the contractor. In this example, there are two sponsors who have throughput contracts with the processing company. The processing company enters into a loan and lease agreement whereby the throughput contract is assigned to a security trustee. The payments due under the throughput contract are paid to the trustee, who uses these to service the debt, any excess cash being paid to the processing company. The throughput contracts signed by Company A and Company B must be disclosed in the companies' consolidated balance sheets.

4.9 Case study: Manchester International Airport – Funding of Terminal 2[2]

The first part of the case study explains the general approach of the British Airports Authority (BAA) to funding construction projects after privatization, and describes the various types of funding facilities available and when they should be used. The second part of the case study describes the specific funding approach used for Terminal 2. BAA owns several airports both in the United Kingdom and abroad, but is not directly involved with the construction industry. Although the case study involves a British domestic company, it can easily be applied to the international situation, so long as other factors such as foreign exchange, international capital markets and local conditions are taken into account.

Figure 4.5. Throughput contract

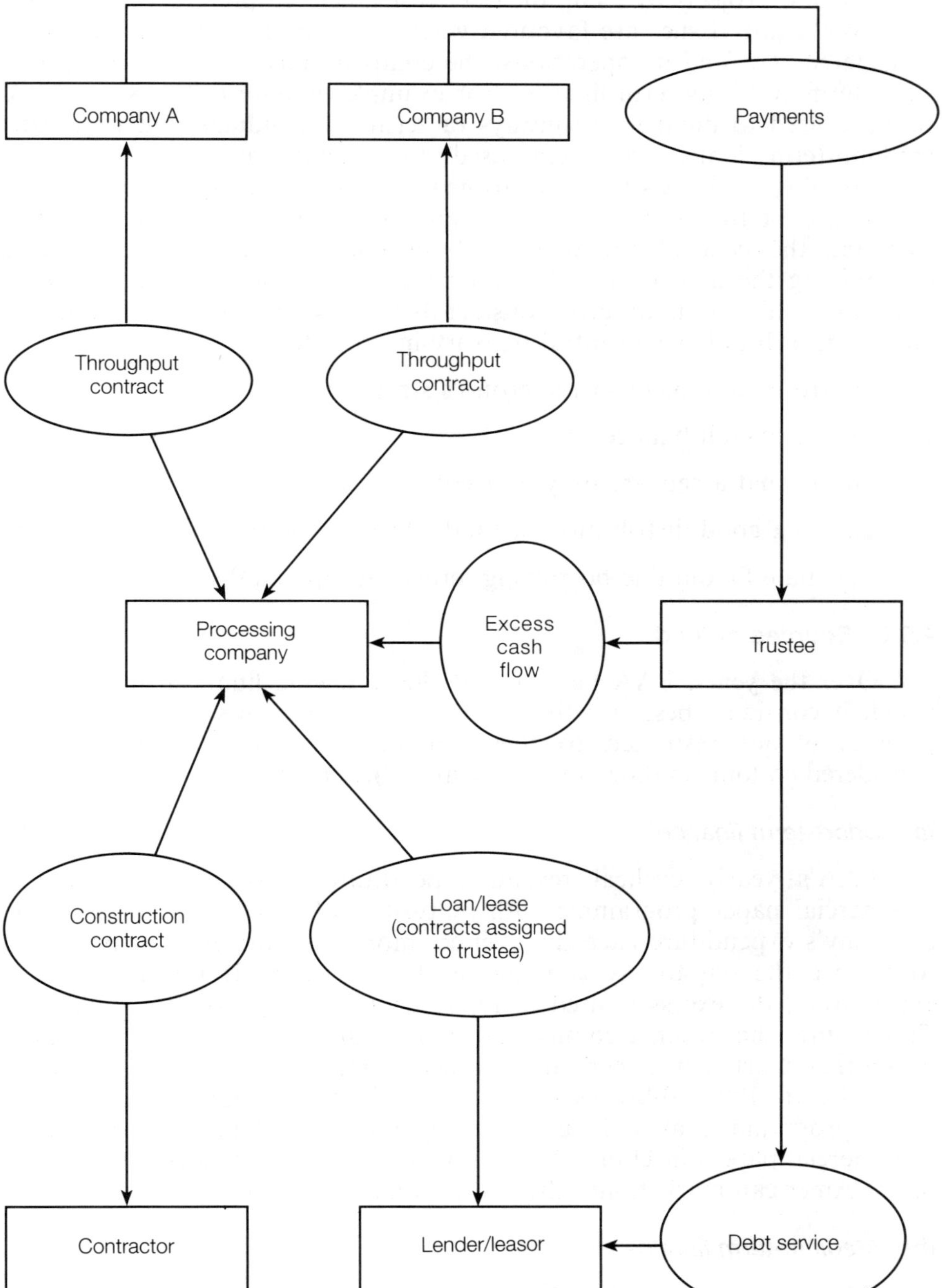

4.9.1 Funding policy

BAA's recent funding policy has been to finance major British construction projects by using the debt instrument in preference to equity. However, equity issues are favoured when international projects are being considered. To fund its operations, the company has to make use of both short-term and long-term finance. For example, long-term loans at a fixed rate are used to build new runways or terminal buildings, but fixed-rate, medium-term loans are often used for smaller projects. It is the responsibility of BAA's treasurer to negotiate with the capital market and obtain a debt that is best for the company in general. This will involve assessing the cost of the debt in both financial and tax terms, and determining the associated risks and possible remedies. BAA's corporate philosophy has been to give considerable autonomy to its subsidiaries. However, it does have a central treasury in order to:

- control and implement the group's strategy;
- reduce overall finance costs;
- ensure that assets are fully utilized;
- ensure a good distribution of funds through the regional offices; and
- negotiate favourable borrowing terms with the banks.

4.9.2 Sources of funds

Over the years, BAA has selected the following financial instruments which it considers best suit the company's requirements. The funding of projects is not restricted to these sources, and new approaches are considered so long as they are beneficial to the company.

(a) Short-term finance

BAA's yearly cyclical revenue/expenditure is balanced through a commercial paper programme with maturities of less than a year. If the company's expenditure exceeds revenue, money is borrowed (short term) to finance the day-to-day running of the company. If revenue exceeds expenditure, the excess is used to repay debts, or pay taxes or dividends. This ensures that when a commercial paper matures it can either be repaid or another short-term paper can be issued to mature when more revenue is available. In 1992, BAA was tied into a US$ 200 million commercial paper programme, as well as a sterling commercial paper programme. Commercial papers in United States dollars are attractive owing to the size of the American market, and the tax allowances that are available.

(b) Medium-term finance

BAA uses the multi-option facility (MOF) as a backup to its commercial paper programme as this tends to be for longer interest periods than commercial paper, but most banks consider it to be high-risk lending and thus impose covenants which ensure some financial control over the company. Owing to its unpopularity with the banks, the MOF has tended to disappear from the capital market. BAA's current MOF comprises 14 underwriting banks and 12 tender banks. BAA's financial requirements are submitted and the lowest bids produced by the tender banks are accepted. The 14 underwriting banks are obliged to provide any shortfall in funds available.

(c) Long-term finance

Long-term debt instruments are used to finance all BAA's infrastructure and construction projects. The main instruments favoured by the company tend to be unsecured loan stock and the sterling Eurobond.

(d) Unsecured loans

As a matter of policy, all BAA's borrowing is obtained on an unsecured basis. The company is fortunate enough to have borrowed £350 million from the European Investment Bank (EIB) in Luxembourg at a non-profit-making interest rate. The EIB has become the company's major lender and often provides up to half the cost of individual construction projects.

(e) Sterling Eurobond

A US$150 million, 25-year sterling Eurobond has recently been issued by the company at an all-in rate of 12 per cent per year. So far, this has been considered to be a good rate although the situation may change over the next 25 years as compared with gilts.

4.9.3 Funding of Terminal 2

Manchester International Airport (MA) plc is a public company owned and operated by BAA. This case study discusses the options that were presented by Norman Renfrew, Director of Finance at Manchester International Airport since 1987, and were considered as part of the feasibility study.

(a) Equity capital

Prior to the construction of Terminal 2, Manchester Airport was considered to be mainly a national airport with international aspirations. In order to fund the planned expansion, the shareholders considered that, as the airport was a regional asset, a public share issue was a possible option. However, there were fears that a future government may well insist on airport privatization, with the consequence of increased private sector participation in both the capital and the management of the airport. Thus, this option was rejected by the shareholders at a very early stage.

(b) Private sector funds

The Treasury rules governing the use of private sector funds are defined in the United Kingdom Airport Act as follows:

> Decisions to provide funds for investment should be taken under conditions of fair competition of private sector borrowers; any links with the rest of the public sector, government guarantees or commitments, or monopoly power should not result in a scheme offering investors a degree of security significantly greater than that available on private sector projects. Such projects should yield benefits of improved efficiency and profit from the additional investment commensurate with the cost of raising risk capital from financial markets.

Under private sector funding, the main financial techniques considered were Build-Own-Operate, joint ventures and lease-leaseback schemes, as discussed below.

(c) Build-Own-Operate

The possibility of financing the project on a Build-Own-Operate (BOO) basis was investigated. A private sector company would have been established to construct and run Terminal 2 on behalf of MA plc, and in return the company would collect revenues on a long-term concessional basis. Thus, the project's risks would be transferred to the new company. At the end of a pre-agreed period, Terminal 2 would be returned to MA plc on a BOO basis. The BOO option was rejected because MA plc did not want to lose control of the operation, and because of the problems with the Treasury Rules which could result in the Treasury rejecting such a scheme.

(d) Joint ventures

A joint venture between MA plc and the airlines was considered to be an option, but this was also rejected. It was found to be too difficult to draw together a group of airlines to finance such a project jointly, and the capacity of Terminal 2 was too large for any single airline. In addition, MA plc wanted to have the flexibility to move airlines between terminals to suit the airport's capacity.

(e) Lease-leaseback

The structure envisaged for the lease-leaseback contract was built around the following construction and operation stages, as explained below:

- MA plc to grant a 50-year ground lease free of charge to a leasing company.
- The leasing company to construct Terminal 2 from its own finances.
- MA plc to have control of both the design and construction of Terminal 2.

- MA to leaseback completed works for a period of up to 25 years and operate the airport.

The proposed lease-leaseback scheme complied with Treasury rules and the strategies involved were acceptable to MA plc. However, it was rejected because the overall cost of financing the operation exceeded the cost of borrowing the money. In addition, the consortium of banks requested a letter from the Secretary of State saying that he would not intervene in the process, a request which MA was not able to comply with.

(f) Public sector financing

Once all the possible sources of private sector funding had been considered and rejected, MA plc had to resort to the use of government funds. However, there was still some concern among the shareholders of the company regarding government intervention and the need to obtain funds on an annual basis.

4.10 Summary

An important feature of the 1980s was the selling off of many nationalized industries by Western governments. This helped to reduce government borrowing and shifted the financial risk into the private sector. The same philosophy has been applied to new construction works, with the Build-Operate-Transfer mechanism (described in Chapter 9) being used to facilitate the construction of many new projects. This chapter discusses some of the important issues that relate to commercial sources of finance, in particular the various types of guarantee and contract that are available. There have been many new developments with respect to guarantees and contracts available, the correct combination of which can reduce the risk associated with finance provided from commercial sources.

Notes

[1] Senior debts have claims on the assets of the borrower which rank ahead of all other debts in case of liquidation. Junior debts are the opposite of senior debts and, in case of liquidation, cannot be repaid until payments on all other obligations have been made.

[2] This case study is based on a presentation made at Loughborough University of Technology, in April 1991, by Fanya Goldwin, Corporate Treasurer of British Airports Authority plc.

References

Andrews, S. 1987. "Project finance: Can Eurotunnel dig itself out of a hole?", in *Institutional Investor* (New York), Vol. 21, No. 11, Nov., pp. 264-272.

Ball, J. 1988. "Project financing: Understanding the mechanism", in *Euromoney* (London), Project Financing Supplement, Aug., pp. 8-10, 15.

Barrett, M. 1986. "Project finance develops new risks", in *Euromoney*, Oct., pp. 73-85.

Beechers, L. 1981. "Contract guarantees and international bonding practices", in *Trade financing* (London, Euromoney Publications), pp. 149-171.

Carnevale, F. 1989. "Project finance: A model for all comers", in *Euromoney*, Apr., pp. 91-92.

Chabotar, K. J. 1985. "Financing alternatives for prison and jail construction", in *Government Finance Review* (Chicago), Vol. 1, No. 3, Aug., pp. 7-13.

Dean, P. N. 1989. "Accounting for development projects: The issues", in *Financial Accountability and Management* (Oxford, Oxfordshire), 5(3), Autumn, pp. 135-148.

Essimi, F. H. 1991. *Financing international construction projects,* unpublished M.Sc dissertation (Loughborough, Loughborough University).

Ferrigno, J. W., 1988. "Project financing: Public and private–It's mutual", in *Euromoney*, Project Financing Supplement, Aug. 1988, pp. 50-51.

Hill, M. 1986. "Co-financing with the World Bank", in *Journal of Commercial Bank Lending* (Philadelphia, Pennsylvania), July.

Hoffman, S. L. 1989. "Project financing: Loans based on cash flow and contracts", in *Commercial Lending Review* (Boston, Massachusetts), Vol. 4, No. 4, Fall, pp. 18-30.

Nevitt, P. 1989. *Project financing* (London, Euromoney Publications, 4th ed.).

Nisse, J. 1988. "Trade finance: Survival of the strongest", in *Banker* (London), Vol. 137, No. 743, Jan., pp. 71-72.

Starkie, D. 1991. "Airport investment: The monopolies and mergers. Commission Report on BAA Plc's South-East airports", in *Journal of Transport Economics and Policy* (Bath, Avon), Vol. 25, No. 3, Sep., pp. 303-306.

Yates, J. K. 1990. "Innovative construction financing techniques", in *Cost Engineering* (Morgantown, West Virginia), Vol. 32, No. 1, Jan., pp. 7-12.

PROJECT FINANCING PROCEDURES OF DEVELOPMENT BANKS

5

5.1 Development projects

Baum and Tolbert (1985b, p. 333) defined a development project as: "a discrete package of investments, policies, and institutional and other actions designed to achieve a specific objective (or set of objectives) within a designated period".

In addition, a development project is often an essential part of a country's development plan. In developed countries similar projects may well go under different names, for example, "infrastructure investment", the difference being that development investment is made specifically to increase capacity, whether it be through structures, organizations or methods. It is wrong to consider development projects purely as physical investments. The objective of any development project should be to improve both physical and non-physical aspects (for example, skills acquisition).

5.2 Sources of finance

Development projects can be financed from several sources. In wealthy countries, such projects can be funded internally from a single source (i.e. the government budget). However, this is not possible in most developing countries, as government funds are often inadequate. If this is the case, development projects will have to be funded mainly from external sources. These external sources often prefer to finance only part of the project, while the remainder will have to be financed internally. Although most foreign-aided projects allocate 100 per cent of foreign expenditure to the external source of finance, there are four main types of transaction that can occur. These are listed below:

- the donor or lender pays for local expenditure;
- the donor or lender pays for foreign expenditure;
- the recipient country pays for local expenditure; and
- the recipient country pays for foreign expenditure.

Figure 5.1. International development agencies

Global
International Bank for Reconstruction and Development (IBRD), World Bank, Washington, DC
International Development Association (IDA), World Bank, Washington, DC
International Finance Corporation (IFC), World Bank, Washington, DC
Multilateral Investment Guarantee Agency (MIGA), World Bank, Washington, DC
International Fund for Agricultural Development (IFAD), Rome
Regional
Abu Dhabi Fund for Arab Economic Development (ADFAED), Abu Dhabi
African Development Bank (AfDB), Abidjan
Arab Bank for Economic Development in Africa (ABEDA), Khartoum
Arab Fund for Economic and Social Development (AFESD), Kuwait
Asian Development Bank (AsDB), Manila
Caribbean Development Bank (CDB), Barbados
European Development Fund (EDF), Brussels
European Investment Bank (EIB), Luxembourg
Inter-American Development Bank (IADB), Washington, DC

There are many international development agencies which provide extensive funding. Some of these are presented in figure 5.1. In most cases, external donors will set a limit on the funds made available. If project costs are higher than expected, resulting in the donor's limit being exceeded, a considerable burden will be placed on the local economy. If several sources are used to finance a project, there are additional problems that may have to be overcome, four of which are detailed below:

- Each source (donor/lender) has a legitimate interest in monitoring the project and thus several accountability reports may be required.
- Accounts will have to be verified through project audits. Very often, one auditor is responsible for the local component, while another looks after the international element. If this is the situation, the jurisdiction of these auditors may not be clear and the local auditor may not be able to review the external element of projects funded in his or her own country.
- As with most construction projects, the executing agency spends money on the project and then claims it back from the sponsors.

Failure to claim back expenditure is uncommon, but extensive delays in submitting claims do occur. This results in additional costs due to both the time-value of money, and any short-term foreign exchange deals that are used to overcome the delay.

- Project expenditure may have to be made in several currencies, and it is difficult to decide how these should be built into the project costs and when they should be presented as a single currency.

5.3 Development banks

Development banks can be defined in a variety of ways and in practice have little in common apart from the basic concept. There are many types of ownership, objectives, sources of finance, policies and degree of government involvement that can be adopted. A development bank is considered by the United Nations to be an institution concerned primarily with long-term loan capital.

However, this vague definition can be applied to most banks. In practice, the main objective of a development bank must be to speed up the development process. In most cases, these institutions have been established by governments mainly to encourage growth within the private sector of their own economy. International development banks can be characterized on a geographical basis, i.e. as national, regional (e.g. African Development Bank – AfDB; Asian Development Bank – AsDB) or universalist (World Bank group) (figure 5.1). The ownership and membership of regional development banks is not restricted to the regional States, hence entities other than States can also be included.

The objectives of most regional development banks are generally restricted to economic development and integration. Some, such as the AfDB, are also concerned with the encouragement of social progress. In order to achieve these objectives, sections are often established within the bank which have one or more of the following functions:

- to facilitate economic construction or reconstruction;
- to help development planning and programming;
- to cooperate with other institutions with similar goals;
- to provide funds for capital investment;
- to encourage and promote both public and private investment from abroad; and
- to generate international capital flow.

Each region has its own problems and regional development banks can produce solutions to encourage economic development, but these have to be restricted to the individual bank's scope of activities. A comparison of the approaches taken towards the construction industry by the different regional development banks would be interesting, but difficult to achieve without full disclosure from the institutions concerned. The policies and, where possible, the procedures adopted by three different regional development banks and by the World Bank are highlighted in the following sections.

5.4 African Development Bank Group (AfDB Group)

The African Development Bank Group comprises:

- the African Development Bank (AfDB);
- the African Development Fund (AfDF), i.e. the soft loan window to member countries with GNP per capita below US$510;
- the SIFIDA Investment Company; and
- the Nigerian Trust Fund (NTF).

The AfDB Group provides both human and capital resources for the economic and social development of 51 African countries. In addition to project/programme lending, it also co-finances projects with the World Bank Group, the European Union (EU) and many other leading agencies.

Notes on the AfDB:

- Established 4 August 1963 in Khartoum.
- 51 regional member countries (1993).
- 25 non-regional member countries (1993).
- Loans to poorer countries usually repayable after 40 years with a ten-year grace period.
- Other loans have maturities ranging from twelve to 20 years with a two- to five-year grace period.
- At the start of 1992, AfDB had assets worth US$9.75 billion and capital to the sum of US$3.154 billion.
- The Bank's participation rarely exceeds 50 per cent of project cost.

The project cycle is very similar to that of the World Bank, as discussed in section 5.8. It usually takes between eight months and three years to complete the approval process. Once the Bank has approved a project involving large and complicated contracts for construction services, it normally requires prequalification for all major supplies. Prequalification is usually based upon:

- financial stability;
- previous experience and performance on similar projects;
- knowledge of local conditions; and

- capabilities with respect to personnel, equipment and plant.

  Contractors who have prequalified are invited to tender and the contract is awarded to the lowest evaluated bidder, defined as being "the bid price which appears to be the most economic after evaluating all terms and conditions of the bid as well as the qualifications of the bidder".

  The Bank also offers many consultancy services which include the following:

- preparation of pre-investment studies;
- detailed design of projects;
- supervision and control of projects; and
- provision of technical assistance.

5.5 Asian Development Bank (AsDB)

The Asian Development Bank encourages economic growth and co-operation in Asia and the Pacific, and also assists in the economic development of its member countries. It comprises the bank itself and its affiliates such as the Asian Development Fund (AsDF), which is the soft loan window providing loans with a 1 per cent per annum service charge, with repayments extending over 40 years including a grace period of ten years.

Notes on the AsDB:

- Established 22 August 1966 in Manila.
- Owned by 36 Asian countries plus 16 countries from Europe and North America.
- Since 1987, most of the AsDB activities have been directed towards economic programme lending.
- At the start of 1993 the AsDB had assets worth US$22.16 billion and capital to the sum of US$7.36 billion.
- Prefers predefined projects.

5.5.1 Funding policy

The AsDB has traditionally been oriented towards the public sector, but this has started to change and the Bank's association with the private sector has grown. Between 1986 and 1989, 18 private sector projects received a total of US$180.7 million from the Bank. However, the AsDB is still reluctant to provide unguaranteed loans to the private sector.

Programme lending can be defined as "a lending mechanism through which loans are provided not for a specific project but for a particular economic programme such as agriculture diversification, etc".

Most governments seem to favour programme lending rather than project lending because of the flexibility that lies within a particular broadband programme. However, the AsDB prefers predefined projects which its staff understand and can account for. In order to overcome this problem and reach a compromise, the AsDB has started to lend to the private sector with government guarantees, by lending to the national development banks which in turn lend directly to the private sector. In addition, the Bank has used other mechanisms such as venture capital, leasing or co-financing. The power station recently built in Hong Kong on a Build-Operate-Transfer basis and the equity interest position in the Nepalese cotton industry are good examples of the AsDB's new approach.

The activities of the AsDB involve mainly transport or communication systems but, where appropriate and possible, funds are made available for education, health and urban development. It is not usual to identify the construction industry as a separate sector when funds are allocated. However, the construction element of most industrial projects usually makes up around 20 per cent of the loan.

5.6 Inter-American Development Bank (IDB)

The Inter-American Development Bank was established on 8 April 1959 in Washington, DC, to promote the investment of public and private capital in the region for development purposes. It currently has 44 members. The IDB comprises:

- the Bank itself;
- the United States Social Progress Trust Fund; and
- the Fund for Special Operations (i.e. the soft-loan window).

Agriculture, rural development, urban improvement and energy development have in the past been the main focus of the Bank's lending strategy. Of its total lending, 35 per cent has gone towards physical infrastructure involving construction works. Until 1987, lending was traditionally directed towards infrastructure projects, for example, roads, bridges, dams and ports. Since then, there has been a deliberate shift away from infrastructure development towards "basic human needs" such as housing. However, construction work remains an important aspect of the Bank's lending policy, even though the number of major construction projects is decreasing.

5.7 The World Bank Group

The term "The World Bank" is used mainly to refer to both the International Bank for Reconstruction and Development (IBRD) and its affiliate, the International Development Association (IDA). There are two other affiliates to the World Bank, namely the International Finance Corporation (IFC) and the Multilateral Investment Guarantee Agency

(MIGA). These four bodies are collectively known as the World Bank Group (figure 5.2)

Figure 5.2. The World Bank Group structure

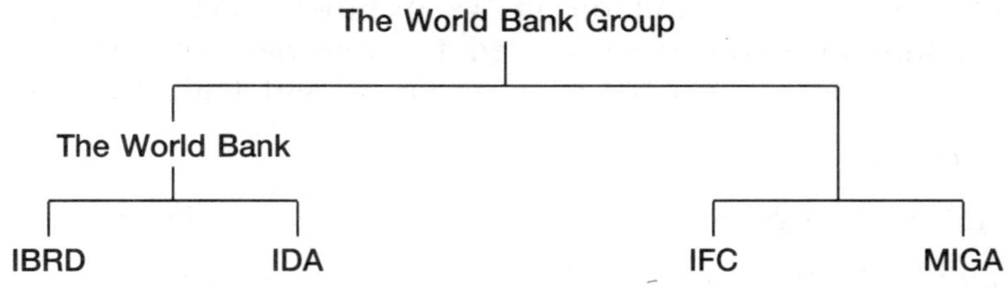

5.7.1 The IBRD

The IBRD's operations are governed by its charter, which spells out certain basic rules.

It aims to benefit member States by:

- promoting economic development by making loans on favourable terms to governments for priority productive projects where finance cannot be obtained on reasonable terms elsewhere;
- providing technical assistance to aid economic development; and
- cooperating with other donors of financial and technical assistance to improve the effectiveness of the International Development Association (see below).

Notes on the IBRD

- Established on 27 December 1945 in Washington, DC.
- Owned by the governments of 151 States.
- Subscribed to by its member States.
- Finances lending operations mainly by borrowing on the world capital markets.
- Substantial resources come from retained earnings and loan repayments.
- Loans are usually repayable over 15 years.
- Grace period for loans of up to five years.
- Directed towards developing countries at more advanced stages of growth.
- Guidelines set out charges to be made on its loans.

5.7.2 The IDA

The IDA aims to promote economic development by providing finance on more flexible terms and bearing less heavily on the recipient country than conventional loans. In the fiscal year 1993, the IDA provided credits amounting to US$6,751 million in 43 member countries. Most funds ($5,921 million in 1993) are provided to countries with an annual per capita income of less than US$635 (in terms of 1991 United States dollars)

Notes on the IDA:

- Established on 24 September 1960 in Washington, DC.
- Open to all IBRD member States (143 States have joined to date).
- Same purposes as the IBRD.
- Primarily for poorer developing countries.
- Balance of payments less important.
- Loans termed as credits.
- IDA credits have ten-year grace periods.
- 35- to 50-year maturities.
- No interest is payable, but there is a service charge of 0.75 per cent.
- Funds obtained through subscriptions, top-ups from richer countries and transfers from IBRD profits.

5.7.3 The IFC

The function of the IFC is to encourage economic development in less developed countries, to stimulate private sector growth and to help mobilize domestic and foreign capital to this end.

Notes on the IFC:

- Established on 24 July 1956 in Washington, DC.
- Separate entity to IBRD.
- Members must be IBRD members.
- Currently includes 143 member States.
- Provides venture capital to productive private enterprises without government guarantees.

5.7.4 The MIGA

The main aim of the MIGA is to encourage flows of investment for productive projects to developing member countries through the mitigation of non-commercial investment barriers. This is achieved by: providing coverage for non-commercial risk; heightening awareness of investment opportunities; enhancing relationships between host governments and foreign investors; and consulting or negotiating on conditions for foreign investment.

Notes on the MIGA:

- Established on 12 April 1988 in Washington, DC.
- 67 member States.
- Offers guarantees against non-commercial risks.
- Advises developing member governments on design and implementation of policies relating to foreign investments.
- Sponsors a dialogue between international businesses and host governments.

5.7.5 Types of project support by the World Bank

Projects supported by the World Bank include the construction of dams, roads, bridges, ports and many other large infrastructure projects. However, there has recently been a significant trend towards smaller projects in the agricultural and social sectors. In the fiscal year ending 30 June, 1991, the World Bank provided US$27.7 billion worth of IBRD loans and IDA credits. Since most projects only receive approximately 35 per cent of the total cost from the World Bank, projects totalling US$60 billion are made possible each year through the support of the Bank. The aid is used to purchase goods, equipment, consultancy services and construction of the works required to complete specific projects.

5.8 The World Bank approach to the project cycle

Most projects and executing agencies have their own characteristics based upon needs that have to be satisfied. Consequently, the final financial package that is used to service the project is often unique to that project. In arriving at the final package, several routes will have to be explored and considerable backtracking may be required. Each funding agency, including the development banks, has procedures and policies that must be followed. Although most development banks have a mandate or charter that governs their actions, and from which its policies evolve, its procedures and policies do change with time. The controlling factors are therefore the policies of the bank, but flexibility may well exist within its procedures. This flexibility is required because of the varied nature of the projects encountered.

This approach provides the executing agency with a clear picture of the work which it has to perform, and indicates when the development agencies should be approached and the type of information that should be presented. The whole process is one of negotiation and should therefore be based upon good communications between the various parties involved, as well as comprehensive documentation until project completion. Rather than describe these procedures one by one, the remainder of this section discusses the various stages of a development project and highlights the main requirements of the development banks (and other investors/sponsors) in each of them.

5.8.1 The project cycle

The definition of a development project was presented earlier. It is possible to break down a project into various stages, often referred to as "the project cycle". This term emphasizes that a close link should exist between the various stages of a project, and that the final stages should provide the basis for further projects. There are several terms that can be used to define the different stages of a project cycle. According to Baum and Tolbert (1985b, pp. 334-335), the World Bank uses the following:

- **Identification.** The first phase of the cycle is to identify ideas that appear to represent a high-priority use of the country's resources to achieve an important development objective. Such project ideas should meet an initial test of feasibility; that is, there should be some assurance that technical and institutional solutions – at costs commensurate with the expected benefits – will be found and suitable policies adopted.

- **Preparation.** Once a project idea has passed the identification "test", it must be advanced to the point at which a firm decision can be made whether or not to proceed with it. This requires a progressive refinement of the design of the project in all its dimensions – technical, economic, financial, social, institutional, and so on.

- **Appraisal.** Before approving a loan, external agencies normally require a formal process of appraisal to assess the overall soundness of the project and its readiness for implementation. For an internally generated and financed investment, the extent of formal appraisal varies widely in accordance with government practice. Some explicit appraisal, however, is a necessary, or at least a desirable, part of the decision-making process before funds are committed.

- **Implementation.** The implementation stage covers the actual development or construction of the project, up to the point at which it becomes fully operational. It includes monitoring of all aspects of the work or activity as it proceeds, and supervision by "on-site" agencies within the country or by external lenders.

- **Evaluation.** The *ex post* evaluation of a completed project seeks to determine whether the objectives have been achieved and to draw lessons from experience with the project that can be applied to similar projects in the future. Although some lending agencies such as the World Bank routinely require an *ex post* evaluation of all projects that they finance, few developing countries have established a comprehensive system for evaluating the results of their project investment portfolio.

This cycle is presented in figure 5.3.

5.8.2 Project identification

Although large capital projects are selected in order to satisfy effective demand, the process should involve the consideration of other economic, technical, managerial and financial issues. Project identification stems from the translation of national development plans into specific action programmes and investments. In formulating such plans, potential projects will have to be identified and ranked in order of priority. The aims of individual projects must be clearly defined and agreed at an early stage. It is essential that all parties involved are consulted, among them the government, the beneficiaries and the funding agencies. One of the parties involved may be more interested in increasing export earnings, where another may prefer to see a reduction in local unemployment and improved distribution of wealth. It is crucial that full agreement is reached on the project objectives by all parties associated with it. In the past, failure to achieve this has often generated considerable friction, resulting in poor project implementation. Individual governments, with assistance from outside agencies, should therefore take a systematic approach to the identification process, which should include the following steps:

- thoroughly review the borrower's country and highlight under-developed sectors with potential for investment;
- formulate general national development plans identifying potential projects;
- compare the potential for development with other countries;
- identify the potential for development in terms of:
  - extraction of domestic natural resources;
  - processing of domestic natural resources;
  - import reduction;
  - export expansion; and
  - expansion of current industries.
- select specific projects for further consideration;
- assess selected projects on the basis of market, technical, resource and financial considerations; and
- formulate a specific national development plan identifying specific projects, their objectives and an appropriate time-scale.

Once the project objectives have been established, it should be possible to present them in the form of a Project Brief. The World Bank has developed procedures to be followed during the preparation of a Project

Figure 5.3. The World Bank project cycle

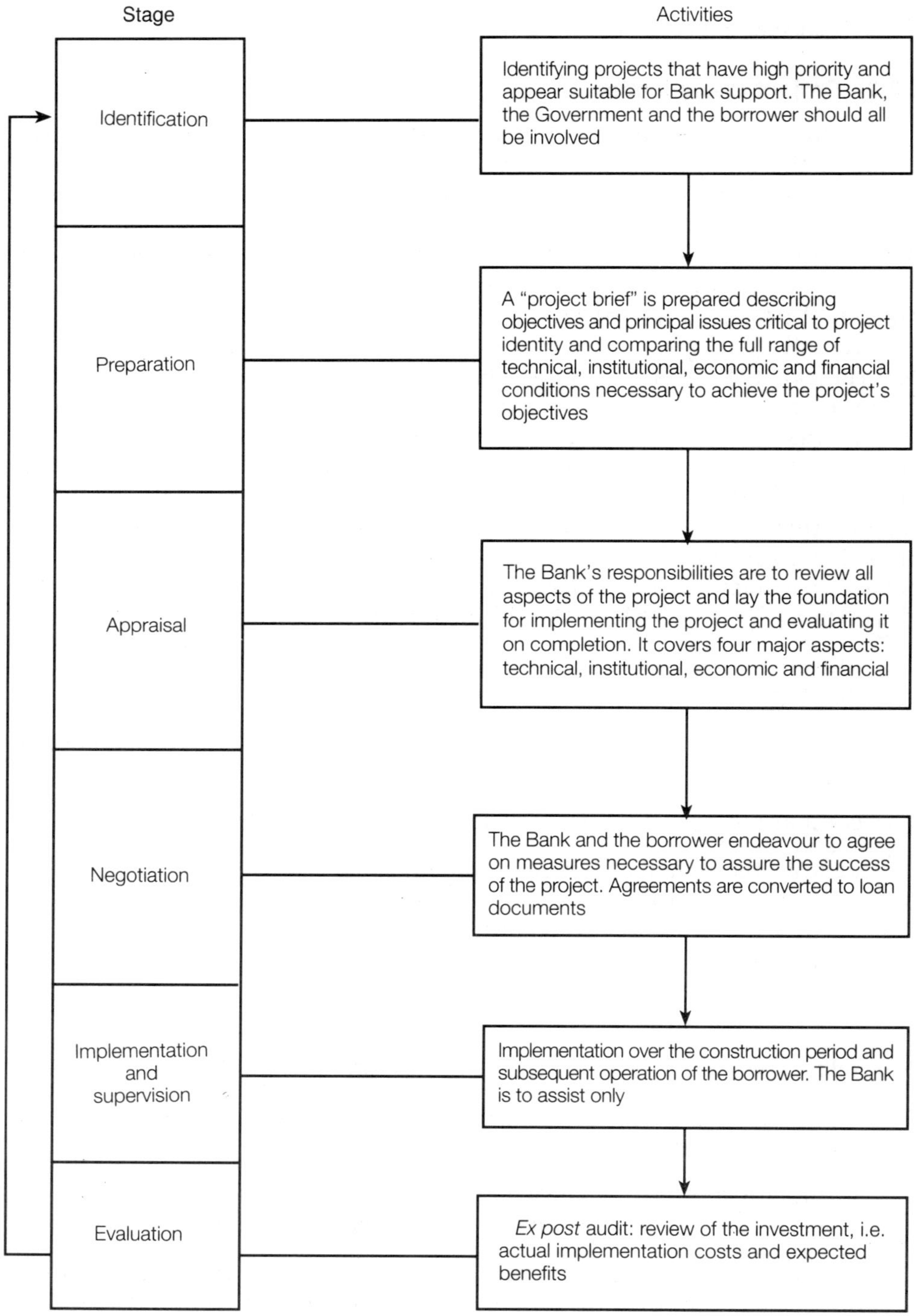

Source: Baum, 1982

Brief. These enable early agreement to be achieved between the World Bank and potential borrowers on the following issues:

- development objectives of the project;
- basic project features;
- alternatives still to be considered;
- institutional, policy and other matters to be considered during preparation, appraisal or implementation; and
- actions required to prepare the project and the main resources to be employed.

The World Bank, along with other multilateral agencies, offers valuable assistance with the project identification phase. In some cases, they have regional offices whose main role is to help identify and prepare projects. Another option is to send identification missions from the head office. United Nations agencies, including the United Nations Educational, Scientific and Cultural Organization (UNESCO), the United Nations Industrial Development Organization (UNIDO), the International Labour Office (ILO), and the Food and Agricultural Organization (FAO) provide specialized services for both project identification and preparation. Another source of financial and technical assistance is the United Nations Development Programme (UNDP), which provides funds for resource surveys or pre-feasibility studies carried out in order to identify suitable projects.

5.8.3 Project preparation

Once a project has been identified, the next stage of the cycle involves project preparation and appraisal. These tasks must be completed before the decision to allocate funds can be made. The process is not straightforward and is often an iterative process. Project objectives and the means of achieving them will have to be refined, several options will have to be considered and a cost-benefit analysis performed.

Project preparation or refinement is not just aimed at assessing whether a project is a good idea and is economically worthwhile, but it should identify the best solution available under the predicted conditions. The main steps of project preparation or refinement are listed below:

- specify production increases expected to result from various phases of the project, and quantify them in various types of output;
- identify project beneficiaries in terms of location, income groups or social characteristics;
- specify the quality and extent of services to be provided;
- examine in detail the technical aspects of the project (e.g. ground conditions);

- develop into action programmes any institutional reforms that have to be achieved; and
- perform a cost-benefit analysis on the remaining alternatives.

The World Bank has tended to adopt a policy of lending at a late stage of project preparation, very often after detailed engineering aspects have been completed and other steps have been initiated. Once the detailed economic and financial analyses have been completed, a valid judgement can be made as to the project's viability before any funds are allocated. However, it is important to realize that funds spent during the preparation stage could produce more concrete results if allocated to the implementation stage.

The cost of project preparation

The cost of a feasibility study varies, but is usually about 5 per cent of the overall project cost. There can be little doubt that good project preparation reduces the possibility of problems during implementation. When the process described above is started, there will be many options available. However, some of these will be eliminated, and when the alternatives have been reduced to a more manageable number, more detailed preparation of the remaining projects can be undertaken in the form of a feasibility study.

The feasibility study

A feasibility study provides both national and international decision makers with a basis for deciding if the project should proceed and identifying the most desirable option available. It is necessary to perform a feasibility study even when a pre-feasibility study has been undertaken. Some agencies prefer to use the term "preparation report", but the objectives are just the same. There are many aspects that form part of a feasibility study, including financial considerations, technical problems, environmental impact and many others. Each of these have to be analysed both separately and in relation to each other.

Sources of technical and financial assistance

There are several organizations that can provide technical and financial assistance. If a project is to be funded from external sources, it is necessary that prospective lenders provide substantial input into the project preparation. If project preparation is undertaken without reference to specific lenders, considerable reworking may be required in order to satisfy the requirements of individual lenders.

Projects funded by the World Bank usually involve the appointment of several consulting services, such as engineering feasibility and design services or construction supervision. These services should be procured in accordance with the procedures outlined in the Bank's publication

Guidelines: Use of consultants by World Bank borrowers and by The World Bank as executing agency (Washington, DC, 1981), which recognize that the selection process should concentrate on the quality of service rather than minimum price. The UNDP is an important source of finance for feasibility studies and similar preparatory work. The World Bank finances this type of work through several routes; for example, preparatory studies for future projects may be included as part of funding for current projects. Other organizations such as the FAO and UNESCO are also valuable sources of funds specifically for preparatory work.

Conducting a feasibility study requires many professional skills. If local staff are used, then details relating to the local environment, social conditions and legal aspects can be included. In some cases, such as agricultural projects, the inclusion of local staff is essential. The development of the ability to produce feasibility studies and undertake similar project work can be beneficial to a developing country and is considered to be a worthwhile objective on its own merit. Some of the more advanced developing countries have established very strong technical departments which are able to compile complex feasibility studies.

5.8.4 Project appraisal

Agencies such as the World Bank, the FAO and UNIDO have produced checklists of topics that should be included in feasibility studies or project preparation reports. Some of these checklists are extensive and have originated from previous projects. These should be used only as a reminder or guide and should not be totally relied upon. The preparation of a feasibility report is a creative process and should not be confined to areas contained in available checklists. However, if these checklists are used as intended, the chances of a favourable response to the proposal from the lenders are increased.

The World Bank relies mainly on its regular staff or appraisal missions for its own appraisal of each project. Project appraisal reports have to pass through an elaborate review procedure before the Bank will enter into negotiations with the borrower and before the application is forwarded to the Board of Executive Directors for their consideration. The current demand for World Bank funds exceeds supply; consequently, additional funds have to be sought from other sources, for example, regional development banks, commercial banks or official government agencies. In addition to generating additional capital, this process enables the risks associated with individual projects to be shared and is known as "co-financing". Where co-financiers adopt different procurement methods, the main difference usually being the countries eligible to supply goods and services, each co-financier can finance individual contracts based on its own procedures; this process is referred to as "parallel financing". When agencies have identical procedures, a common funding arrangement can be adopted, i.e. "joint financing".

5.8.5 Project implementation

The importance of the early stages of the project cycle has already been emphasized. However, project implementation is the critical part of the project, as it is here that the earlier plans, designs and analyses are to be tested in the harsh light of reality. Although the responsibility for project design and implementation lies with the borrower, the World Bank is deeply involved throughout the process. Implementation begins when resources are committed to an investment. The start of the implementation stage for certain projects, such as roads or buildings, is easily identified, but projects comprising a variety of activities may not have a clear commencement date.

To have received World Bank financing, project proposals will have detailed the procurement procedures in line with the Bank's publication, *Guidelines for procurement under World Bank loans and IDA credits* (Washington, DC, 1985). These procedures form part of the Loan Agreement. The borrower plans and schedules the procurement, prepares bidding documents, evaluates the bids and awards the contracts. The Bank will pay out its funds only if the terms of the Loan Agreement have been followed, and this includes the procurement procedures.

Planning the implementation stage of a project has only recently started to receive the attention it deserves. Previously, work done during the earlier stages concentrated upon establishing alternatives and assessing their feasibility, along with the preparation of action programmes.

There is a definite need to plan in clear detail the implementation stages of the project, ensuring that adequate account is taken of the institutions which will execute and operate them. Three main aspects of the implementation stage must be given consideration if the project is to be successfully implemented. These involve: (a) selecting the implementation unit; (b) supervising project implementation; and (c) establishing the co-ordinating mechanisms. These are discussed in detail, along with (d) monitoring and evaluation of project implementation.

(a) Selecting the implementation unit

Most developing countries have few resources that can organize and manage large projects. If complex designs are produced, then the demands on these resources will increase and the project implementation will be subjected to unnecessary problems. An imbalance between the capabilities of the implementation agency and the project's size or technical requirements has in the past often proved to be a major cause of problems during project implementation. Some projects may be considerably larger than anything that the implementation agency has previously experienced. This situation usually occurs in poorer countries or where the first loans are made to a specific agency.

The implementation of smaller projects may be incorporated within an agency's routine programme. For example, the construction of a new road could be implemented within the highway department's programme. However, some projects may be, by comparison, very large or may come under the remit of several agencies. If this is the case, a special project unit may be set up within the agency to oversee the project implementation. On previous projects, this approach has often been a source of conflict, and hence the practice of the World Bank has been to create project units detached from the established agencies and operate them under a different set of rules. These units will have their own administrative procedures regarding the use of funds and staffing regulations.

While this approach lends itself to improved project implementation, it does have its drawbacks. There are no long-term institutional benefits, as the units' rules and methods are not easily transferred. Existing agencies may suffer, as their staff may leave to join these special units. Therefore, this approach should only be taken when the benefits to be obtained through improved management of the investment are greater than the disadvantages to existing organizations.

(b) Supervising project implementation

The implementation of every project is monitored by World Bank staff through correspondence and field visits. A considerable amount of time has to be spent reviewing draft bidding documents and evaluation reports prior to the award of contracts. Some contracts are selected for a review by the Bank before loan disbursement. If the borrower awards a contract that does not conform to the Loan Agreement, no disbursement is made and that part of the loan is cancelled. In the case of a dispute, the Loan Agreement takes precedence under the generally accepted principles of public international law.

The World Bank (and other external agencies) will often have a supervisory role. However, this is not a substitute for the country's own responsibility to ensure that project implementation is performed effectively and efficiently. Supervision may be provided from another unit within the same agency, so long as it is administratively independent of the units doing the work. Alternatively, another agency or outside consultants may be used to oversee project implementation.

(c) Establishing the coordinating mechanisms

It is not unusual to have several agencies involved in the implementation of multi-component projects, such as rural or urban development schemes. However, it is essential that these agencies are able to communicate effectively and can coordinate their actions. In order to achieve this, coordination mechanisms have to be established. There are several approaches that could be taken, but to ensure full coordination it is necessary to:

- clearly define the responsibilities of each participating agency; and
- provide sufficient incentives to encourage them all to contribute to successful project completion.

In the past, three different coordinating mechanisms have generally been adopted:

- establish a special project unit which allocates funds to each of the implementing agencies through a central budget;
- appoint one of the implementing agencies as the lead unit, which the remaining agencies and dispenses some of the funds; and
- appoint a committee with the appropriate authority to coordinate the implementing agencies. The members of this committee must have close contact with the project. In the past, committees comprising ministers or the like have not functioned effectively.

(d) Monitoring and evaluation of project implementation

The World Bank's experience has demonstrated that the monitoring and evaluation of project implementation can provide an early indication of potential problems, and the system adopted should be kept as simple as possible, while following some specific guidelines. In the past, some monitoring systems have proved to be complicated and time-consuming, without providing the results required. The basis of a monitoring system should be a predefined, simple set of indicators (such as early traffic flows). Ongoing project evaluation involves the reassessment of project objectives and the methods being used to achieve them. Some projects last for several years, and over this period many new developments occur and have to be taken into account. Very often, project monitoring and evaluation are added too late during project implementation and are designed without due consideration of the following elements required to produce an effective system:

- management must need and be prepared to use the information produced;
- the data collection must be directed towards the difficulties that have to be overcome during implementation;
- realistic levels of accuracy and cost must be programmed into the data collection; and
- data should be collected early on, in order to provide the basis for future comparisons.

5.8.6 Project evaluation

The *ex post* evaluation is the last part of the project cycle, and is performed once a project has been completed and has started to operate.

The aim of such an evaluation is to take a broad look at the potential benefits of the project once it has been completed, and compare these with the original objectives. If the project is allowed to run for a defined period, the *ex post* evaluation can be based upon actual project costs, operating costs and any early benefits. It should determine whether or not the project was worthwhile and also indicate whether:

- the original objectives were properly defined and achievable;
- appropriate technical choices were made;
- appropriate procurement procedures were adopted;
- the project's target group was correct;
- the local conditions were correctly identified;
- there were excessive cost overruns; and
- project institutions were strengthened.

The World Bank was the first of the multilateral aid organizations to develop a comprehensive *ex post* evaluation methodology and make the findings readily available. Although all the projects funded by the Bank have to produce a completion report, only about half of them are subjected to a detailed audit by the Operations Evaluation Department (OED) (in fact 10 per cent of these will be selected at random). An annual summary of the evaluated projects is produced by the OED staff, and is published as the *Annual Review of Project Performance Audit results,* with country and project omitted to ensure client confidentiality. There are also certain types of project that may be specifically selected for a performance audit. These may include: projects that suggest a poor performance or have had particular problems; projects that are the first of their type in a particular country; or projects that are similar and can thus form part of a comparative study. These procedures were designed to satisfy the Bank's own requirements, but can easily be adopted by other agencies and comprise two phases:

Phase I

Every World Bank assisted project is reviewed by the operational staff within two years of final disbursement. Phase I involves the preparation of a project completion report by the individuals directly involved in the project preparation and implementation.

Phase II

Phase II in the evaluation process is performed by OED staff detached from the operational side of the project. This phase is required to ensure that objectivity is achieved during the evaluation process.

It is inevitable that disagreements will occur between reports and audits produced by the OED staff and the operational staff. A large proportion of these discrepancies can be clarified through discussions. However, the OED staff have the responsibility for the final audit memorandum, although any remaining disagreements should be noted. The borrowers are also included in this evaluation process. As part of the loan agreement, and in addition to other reporting commitments, the borrower has to produce a project completion report.

5.9 Other approaches to the project cycle

Section 5.8 describes the World Bank's approach to the project cycle. The various stages are defined and the associated activities discussed. The cycle is deemed to start with project identification and finish with an *ex post* audit review of the investment, in order to assess actual implementation costs and expected benefits. The process is referred to as a cycle because the results of the *ex post* audit should be fed into subsequent project identifications. This concept of the project cycle is not confined to the World Bank; many other aid agencies take a similar approach. However, there are some differences as to how the project cycle is broken down into the various stages and the activities performed during each stage. To illustrate this point, the approaches taken by the World Bank, the Kreditanstalt für Wiederaufbau (KfW) in Germany and the Asian Development Bank (ADB) are presented in figures 5.3 to 5.5. In addition, the development stages for water supply and sanitation projects are presented in figure 5.6 to show how the World Bank's approach is occasionally modified. The information contained in these diagrams has been extracted from the references given, but has been modified into its current form to illustrate the different approaches taken to the project cycle.

5.10 Summary

In most developing countries, there are many major construction projects that would not have been possible without the financial aid provided by the development banks. This chapter discusses the approaches taken by many of the national, regional and global development banks, and outlines the conditions that have to be satisfied if funds are to be attracted. The project cycle has been discussed at length in order to highlight the various stages which must be considered when either putting a case forward or justifying expenditure once a project has been completed. The project cycle will vary from bank to bank and it is crucial that such differences are recognized if sufficient funds are to be obtained at a reasonable cost.

Figure 5.4. The KfW project cycle

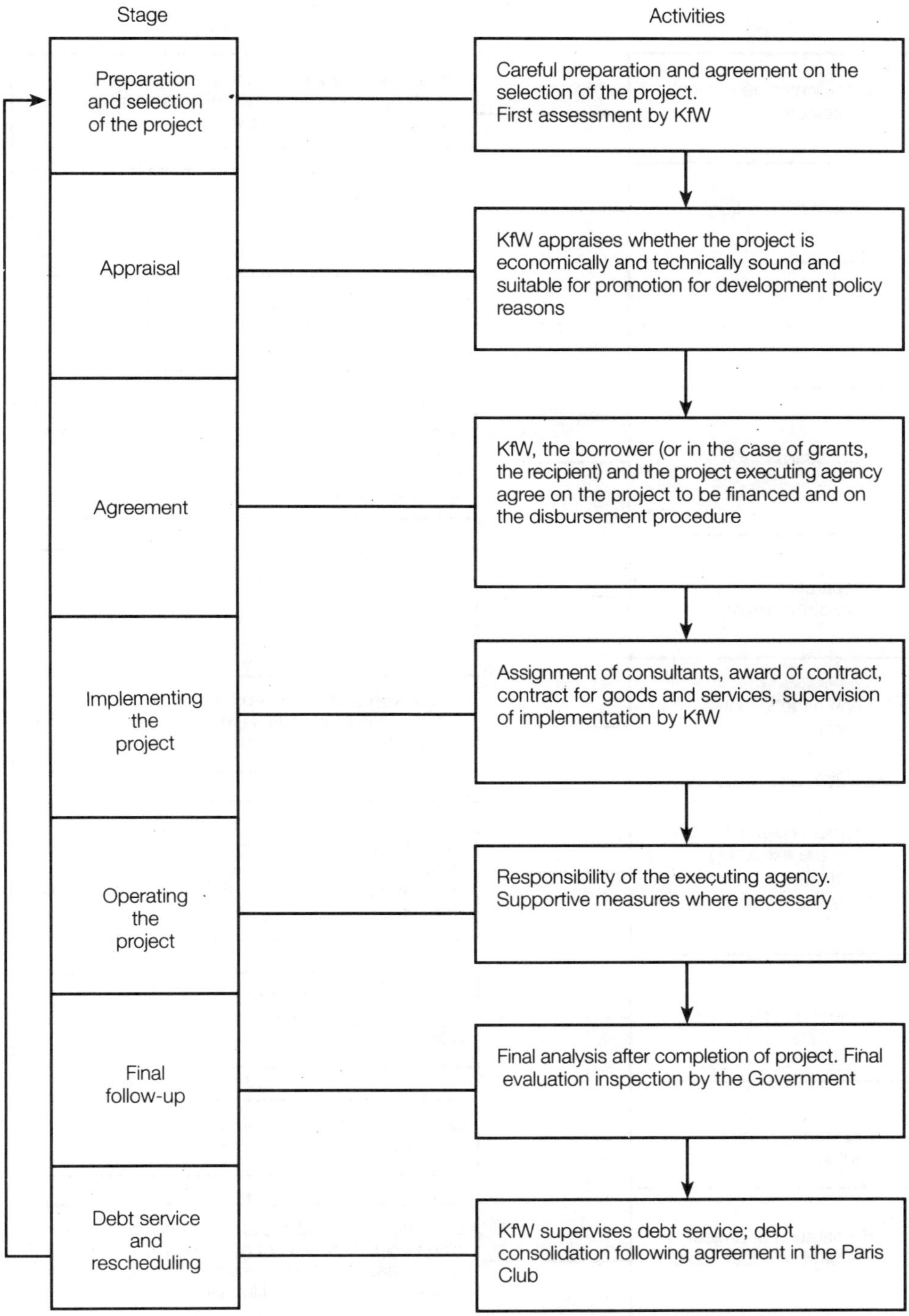

Source: KfW, 1990.

Figure 5.5. The Asian Development Bank project cycle

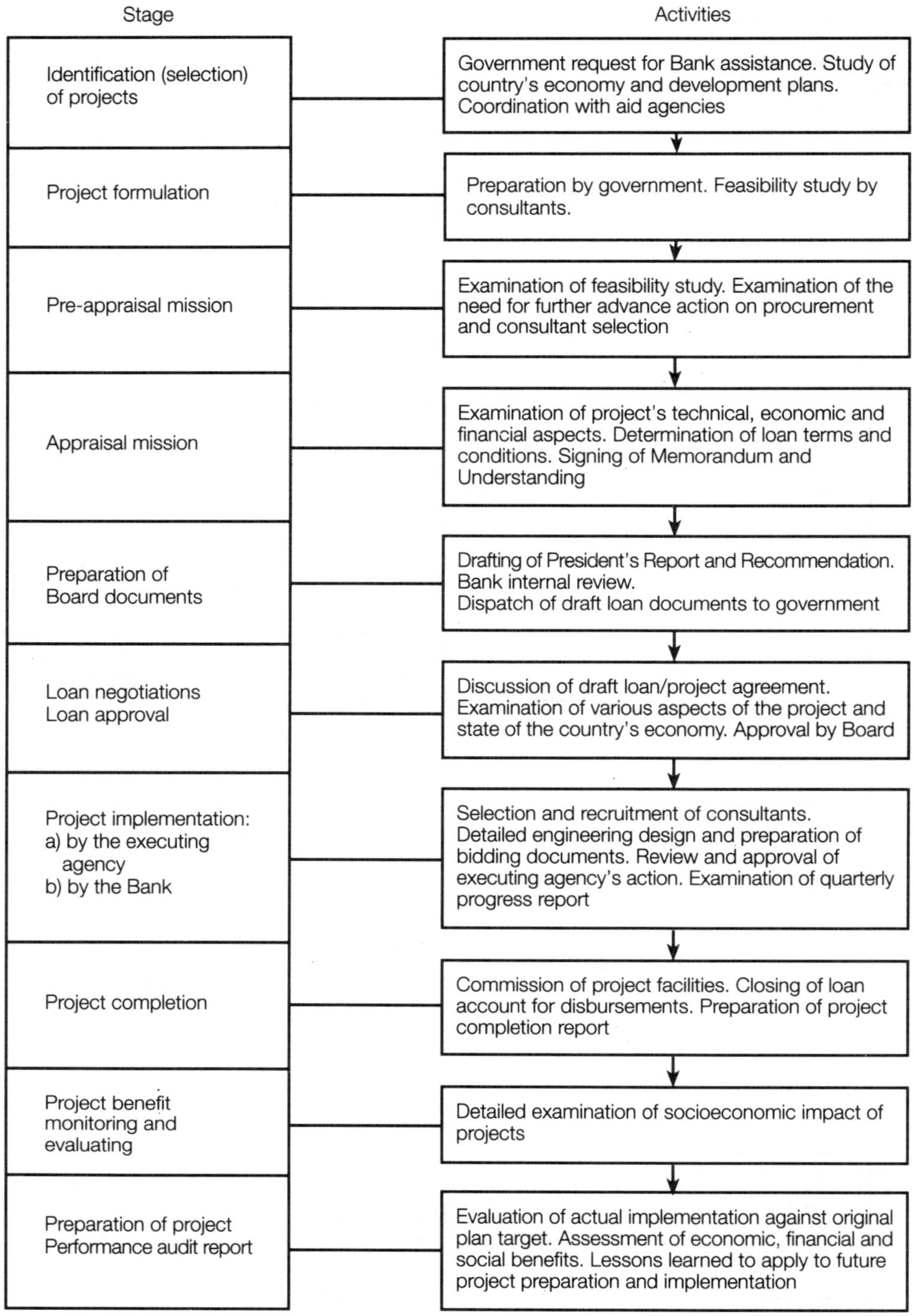

Source: AsDB, 1991.

Figure 5.6 Development stages for water supply and sanitation projects

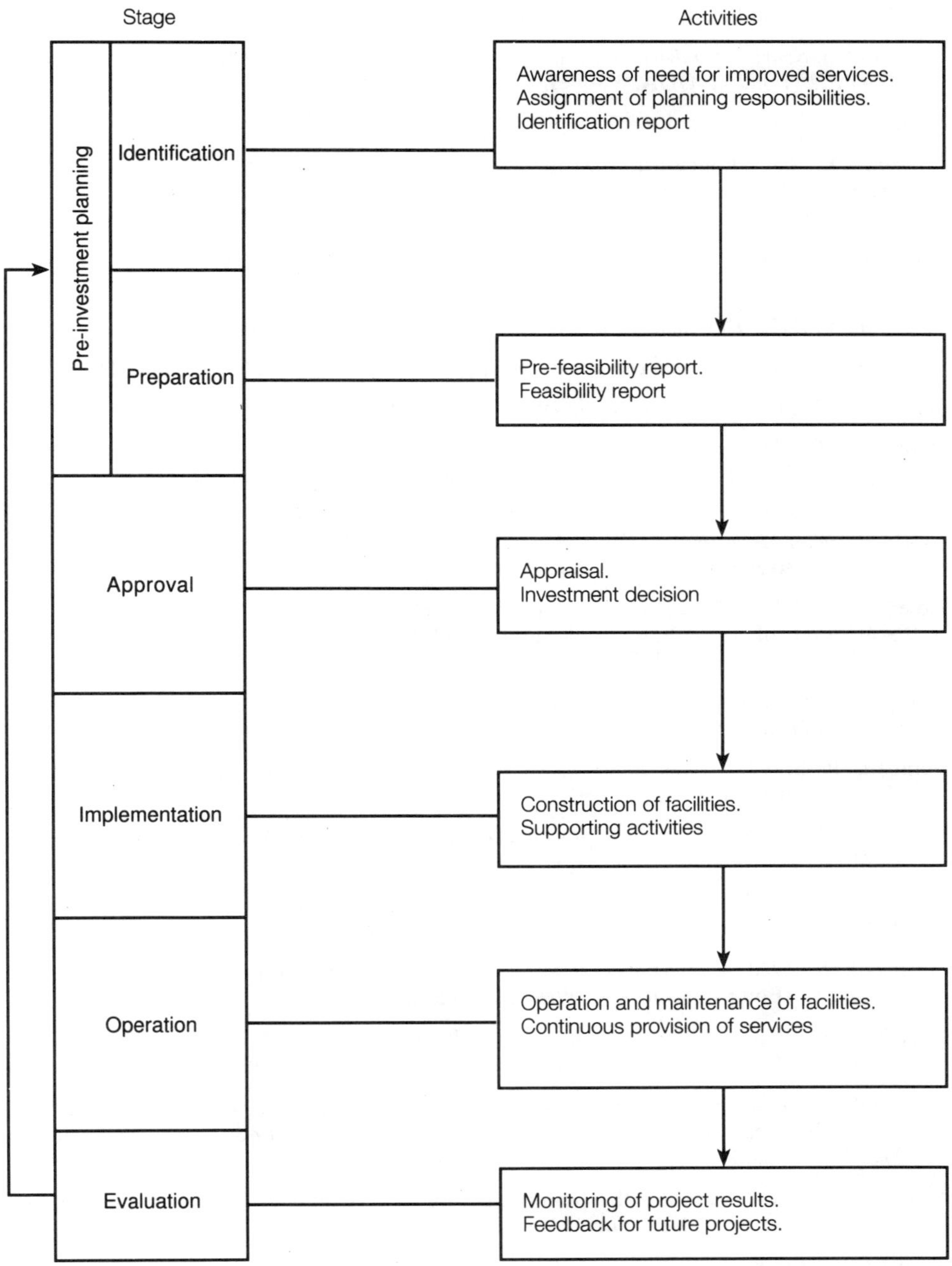

Source: Grover, 1983

References

Ackroyd, P. J. 1991a. *The design and appraisal of bilateral aid proposals*, Office Instruction, GAPE 1 (London, Overseas Development Administration (ODA), 3 June).

——. 1991b. *Sources of advice on the design and appraisal of aid proposals,* Office Instruction, GAPE 2 (London, ODA, 3 June).

African Development Bank (AfDB). 1990. *African Development Fund: In brief.* (Abidjan).

Akroyd, D. 1985. "Policy priorities for agricultural development in Africa with particular reference to agricultural project planning" in *Agricultural Administration* (Barking, Essex), No. 19, pp. 101-113.

Asian Development Bank (AsDB). 1991a. *Asian Development Bank: Basic information* (Manila).

——. 1991b. *Asian Development Bank: Questions and answers* (Manila).

——. 1991c. *Asian Development Bank : What it is: What it does: How it works* (Manila).

Baum, W. C. 1982. *The project cycle* (Washington, DC, World Bank).

Baum, W. C.; Tolbert, S. M. 1985a. "Investing in development–Lessons of World Bank experience", in *Finance and Development* (Washington DC), Vol. 22, No. 4, Dec., pp. 26-36.

——;—. 1985b. *Investing in development – Lessons of World Bank experience* (Oxford, Oxford University Press).

Coleman, G. 1987. "Project planning. Logical framework approach to the monitoring and evaluation of agricultural and rural development projects", in *Project Appraisal* (Guildford, Surrey) Vol. 2, No. 4, Dec., pp. 251-259.

European Investment Bank. 1990. *Annual Report 1990* (Brussels).
——. 1991. *Financial facilities under the Fourth Home Convention* (Brussels).

Feinberg, R. E., et al. 1986. *Between two worlds – The World Bank's next decade.* (Oxford, Transaction Books).

Grenfell, J. 1991. *Business opportunities in projects financed by the World Bank* (Paris, World Bank).

Grover, B. 1983. *Water supply and sanitation projects preparation handbook: Guidelines,* World Bank Technical Paper No 12 (Paris, World Bank).

Hurni, B. S. 1980. *The lending policy of the World Bank in the 1970s: Analysis and evaluation* (Boulder, Colorado, Westview Press).

International Bank for Reconstruction and Development (IBRD), 1972. *World Bank operations: Sectoral programs and policies* (London, Johns Hopkins University Press).

——. 1988. *The International Bank for Reconstruction guidelines: Use of consultants by World Bank borrowers and by the World Bank as an executing agency* (Washington, DC, World Bank).

Jain, P. K. 1989. "Assessing the performance of a development bank", in *Long Range Planning* (Elmsford, New York), Vol. 22, No. 6, pp. 100-106.

Kreditanstalt für Wiederaufbau (KfW). 1990. Cooperation with developing countries. *The procedures followed in the financial cooperation of the Federal Republic of Germany* (Frankfurt).

LeGault, R. 1984. "An exploration into overseas development", in *Real Estate Finance Journal* (Boston, Massachusetts), Fall, pp. 52-56.

Management of international construction projects. 1984, Ch. 1, "Project finance–The multilateral development banks" (London, Thomas Telford).

Syz, J. 1974. *International development banks* (New York, Oceana Publications).

The World Bank. 1980-91. *The World Bank Annual Report* (Washington, DC), various years.

——. 1981. *Guidelines: Use of consultants by World Bank borrowers and by the World Bank as executing agency* (Washington, DC).

——. 1985. *Guidelines for procurement under World Bank loans and IDA credits* (Washington, DC).

——. 1988. *Procurement and co-financing: Methods and procedures* (Washington, DC).

——. 1990. *Guide to international business opportunities* (Washington, DC).

——. 1991. *Bidding for contracts in World Bank financed projects* (Washington, DC).

Union of International Associations. 1993. *Yearbook of international organisations,* Vol. 1, (New York, K.G. Saur Munchen).

United Nations. 1979. *Development banking in the 1980s,* Proceedings of the UNIDO/World Bank Symposium, Zurich, June.

Water Companies Association, United Kingdom. 1991. *Water Aid Annual Review.*

Westlake, M. 1993. "Development Banks: Only the best will do", in *The Banker* (London), May, pp. 23-29.

SOFT LOANS, PREFERENTIAL LOANS AND EXPORT CREDITS

6

6.1 The export credit mechanism

The sale of goods or services to foreign countries is considerably more complex and risky than domestic sales. The main risks that exporters have to face are as follows:

- the buyer may be less well known to the exporter than current clients;
- the exporter may well have to deal with different legal systems, trade customs or languages;
- most buyers prefer to pay for goods only after they have been received and inspected. This can create considerable delays in payment;
- the exporter may have to deal with exchange control regulations;
- changes in exchange rates may well adversely affect the buyer or seller; and
- foreign governments may delay the payment further by temporary restrictions on currency payments.

Export credits are a mixture of insurance and banking mechanisms specifically designed to help exporters overcome the main problems associated with overseas trade. Thus, exporters are encouraged to enter markets that they would find too costly if funded only through commercial sources or bank loans and related insurances. Export credit is therefore a form of home government support for overseas trade. For example, the Export Credits Guarantee Department (ECGD) was set up in the United Kingdom in 1926 as a subdepartment of the Department of Overseas Trade. The role of the ECGD was to provide a facility regularized by the Overseas Trade (Credits and Insurance) Act, 1920. Its main objectives were to reduce unemployment shortly after the First World War and at the same time re-establish trade links which had been cut during that war. However, it was not until 1945, some 35 years later, that a facility specific to the needs of the construction industry was introduced in the form of a "constructional works" guarantee.

Unlike the situation in other countries, the ECGD has never become a direct provider of funds but has offered services such as performance bond cover, pre-shipment finance and cost escalation cover. In 1981, Cecil Parkinson, the then Minister for Trade, summarized the British Government's approach as follows (Royal Institute of International Affairs, 1981, p. 4): "We must support our exporters so that they are competitive; we must match the competition in offering credit, but if we overdo it to the extent that we subsidize inefficient industries, we are simply institutionalizing inefficiency at the taxpayers' expense."

In order to operate in the international market, it has been necessary for all major exporting countries to develop their own export credit facilities. Because of the differences in political and economic systems from one country to another, no two export credit facilities are the same. However, since the problems faced by all countries are very similar, the development of export credit facilities has encountered similar regulations and the same international guidelines. The types of facility provided by the export credit agencies usually fall into two categories: insurance and financing. Export credit can be provided under either a supplier credit agreement, whereby the exporter is financed by a bank and extends credit to the foreign buyer, or a buyer credit agreement, whereby the foreign buyer is provided with finance from a bank in the exporter's country but payment is made directly from the bank to the exporter. Under buyer credit, the foreign buyer is responsible for repayment to the lending bank.

Export credits to foreign buyers may be granted for periods of between two months and ten years, and are usually described as follows:

- Short-term: six months or less;
- Medium-term: between six months and five years; and
- Long-term: over five years.

6.2 Export credit insurance

The major exporting nations have developed "export credit insurance" to help reduce exporting risks. Of most concern is the risk of non-payment for either commercial or political reasons. Both of these risks can be covered by export credit insurance. International contractors can adopt either "comprehensive" or "specific" policies to insure against this type of risk. "Comprehensive" policies are used to cover trade of a repetitive nature, usually with the same client and involving goods which are sold for cash or on a short-term credit basis. Approximately 80 per cent of the ECGD insurance accounts are comprehensive policies. "Specific" policies are to insure non-repetitive contracts requiring substantial financial resources over long durations. Therefore, international construction contractors tend to favour specific policies rather than comprehensive ones. Both types of insurance require the contractor both to pay a premium (usually adjusted to suit the level of risk) and to share some element of the risk, say between 10 and 15 per cent of the contract value.

Specific guarantees were introduced in the United Kingdom in 1954, and proved to be an excellent facility in generating capital at low rates of interest to contractors and exporters. This changed the role of the ECGD from that of a purely credit insurance organization to that of a sophisticated export bank. The ECGD was pushed further into the world of finance when the buyer credit facility was introduced in 1961.

6.3 Export credit financing

Export credit agencies may provide either medium- or long-term credit for payment of goods and services associated with a contract. This facility is known as "export credit financing" and enables exporters to avoid a bottleneck of funds when the client wishes to defer payment. The export credit agency will not necessarily supply the finance, but makes credits available in the form of supplier credit or buyer credit.

Supplier credit allows the client to defer payment but ensures that the supplier receives payment as soon as possible after completion of the contract. The client is effectively provided with credit through the supplier. Promissory notes or bills of exchange should be used to ensure that buyers make the extended payments required. These notes are purchased by the supplier's bank so that payments are made from the buyer to the bank. This process is guaranteed by the export credit agency against default, thus covering this risk.

Buyer credit means that the export credit agency insures the lender (e.g. a commercial bank or the export agency itself) against default on a specific loan arranged between the lender and the buyer. The buyer pays the loan directly to the lender, and the supplier is paid by the lender on delivery of contract completion. The following parties may be involved in buyer credit:

- borrower;
- buyer;
- export credit agency;
- exporter/supplier; and
- exporter's bank.

6.4 Services and cover provided by export credit agencies

Typical services offered by the various export credit agencies are explained below:

- Comprehensive short-term guarantees protect exporters of consumer goods sold on credit not exceeding 180 days.

- Comprehensive extended-terms guarantees are used to supplement comprehensive short-term guarantees by exporters of consumer goods sold on credit exceeding six months but less than five years.
- Comprehensive bank guarantees supplement the holders of comprehensive short-term guarantees, including those with extended terms, who are trading on credit terms not exceeding two years. Guarantees are provided directly to financing banks.
- Comprehensive external trade guarantees protect merchants trading in goods exported from one overseas country to another against the risk of non-payment. Credit terms must not exceed 180 days.
- Services guarantees are intended for organizations providing services to overseas clients. A comprehensive option is available for a recurrent pattern of business, and a specific option for the provision of a one-off service.
- Specific guarantees protect exporters of goods not suitable for comprehensive policies, such as major construction projects.
- Buyer credit guarantees are usually only available on contracts in excess of US$1 million. They enable exporters to be paid on cash terms either on completion or in the form of progress payments by guaranteeing bank loans provided to an overseas buyer.
- Cover for lines of credit protect credit made available to overseas banks by home banks.
- Specific bank guarantees are available to holders of comprehensive extended terms or specific guarantees with contracts based on credit exceeding two years, and provide a guarantee to banks providing finance at preferential export interest rates.
- Cost escalation cover usually protects exporters of capital goods against increases in home costs on contracts over US$10 million and with a manufacturing period of at least two years.

Guarantees provided by the export credit agencies tend to cover from 90 to 95 per cent of losses resulting from the following risks:

- insolvency of the buyer;
- the buyer's failure to pay within six months for goods accepted;
- the buyer's failure to take up goods that have been dispatched (where not caused or justified by the exporter's actions, and where the export credit agency decides that legal action would serve no useful purpose);
- government action which blocks or delays transfer or payment to the exporter from the buyer's country or from a third country through which payment must be made;

- any other action by the buyer's government which prevents performance of the contract in whole or in part;
- political events, economic difficulties, or legislative or administrative measures arising outside the home country, which prevent or delay payment;
- legal discharge of a debt (not being legal discharge under the proper law of the contract);
- war, revolution or similar disturbance in the buyer's country;
- cancellation of the order resulting in non-renewal of an export licence or imposition of new export licensing restrictions.

6.5 Soft loans

Soft loans are long-term, low-interest loans made available to cover the exporter's content of a particular project. The recipient country must be eligible for this type of loan, which is provided by commercial banks in the exporter's country, with support from government agencies. In the United Kingdom, this includes interest support from Aid and Trade Provision funds (see below) and repayment guarantees from the ECGD.

In 1986 soft loan facilities were provided for specific projects in both China and Indonesia. Soft loans are considered on an individual basis, but only where they are clearly going to be more effective than other types of aid such as grants or export credits. Soft loans tend to require higher amounts of input aid, have high administrative costs and absorb larger amounts of limited export guarantee cover allocated to the recipient countries. Therefore, grants or credits, as discussed in Chapter 5, are generally preferred to soft loan facilities.

6.6 The Aid and Trade Provision (ATP)

Although there have been considerable attempts to reduce international trade subsidies, most developed countries still have some form of subsidy guarantee to aid the export of its goods or services. In the United Kingdom, aid funds and fixed-rate export credits are available for some overseas projects through the Aid and Trade Provision (ATP) facility.

The ATP was set up in 1977 to support sound projects of particular industrial and commercial importance to the United Kingdom in specified countries. A separate allocation within the British overseas aid programme has been made for this type of support, but it is discretionary and made available through agreement with Ministers. ATP offers are made to the government of the developing country and are subject to a government-to-government agreement. ATP finance is usually offered with an ECGD Buyer or Supplier Credit Facility. The Department of Trade and Industry (DTI), the Overseas Development Administration (ODA) and the ECGD work in close collaboration and jointly administer the facility.

6.6.1 Criteria for assistance

To be supported by an offer of aid under the ATP, projects must satisfy certain commercial and industrial criteria. They must be both financially and economically viable, and technically sound, before even being considered. Although the principal aim of facilities such as the ATP is to improve the industrial capability of the developing country, they can also benefit the donor country by:

- matching assistance obtained by competitors;
- helping develop new markets;
- securing orders of particular importance to the donor; and
- improving links with developing countries.

All proposals will be assessed against common categories of potential benefits to both the United Kingdom and the recipient country concerned. It is essential that applicants demonstrate the value of the project under the following headings:

- the industrial case;
- the commercial case;
- the soundness of the project;
- the applicant's capabilities; and
- the prospects of success.

The British Overseas Trade Board has published a document, *The Aid and Trade Provision: Guidelines for applicants* (1991), which explains these criteria for assistance in more detail, along with the procedures to follow when applying for ATP support.

6.6.2 ATP support

Most ATP offers can be divided into three main categories. **Specific Matching Offers** are intended to match aid provided to competitors in support of a particular project. If past records indicate that competitors from other countries will secure aid in support of a bid, a **Conditional Matching Offer** may be made. This may be converted to a **Specific Matching Offer** once the ECGD or the ODA have confirmed the level of support available to the competitor. **Initiation Offers** are intended to persuade developing countries to allocate a particular project to the United Kingdom; thus, they are not available in tender competitions where they are seen simply to invite offers from other countries.

There are also two other forms of ATP offer. A **Potential Matching Offer** may be made when there is no firm indication that aid will be provided by other governments, but the possibility is still there. A **No Aid**

Agreement may be made between the governments of competing companies to ensure that they are competing on the same basis.

The exporting contractor must take the initiative in seeking ATP support for a particular project. The route taken depends upon the size and type of project and, in the first instance, contact should be made through the DTI as indicated below:

- if the value of the contract is more than £10 million sterling, contact should be made through the appropriate Project Exporting Division or the Engineering Markets Division; or
- when the value of the contract is less than £10 million, contact should be made through the appropriate Overseas Trade Division country desk.

6.7 Evaluating export credit offers

The project promoter should ensure that invitations are sent to the appropriate export credit agencies of the suppliers of materials and services that stand a good chance of being awarded a contract. These invitations should describe the project and identify the goods and services most likely to be purchased from the suppliers. The replies should indicate the extent of each agency's interest in the project, along with the terms available. The credit terms received from the different agencies can be compared using net present value (NPV) and discounted cash-flow field techniques (Price, 1994). The overall process is illustrated in figure 6.1.

There are considerable advantages of using government-supported finance. However, the following disadvantages should also be considered:

- The cheapest suppliers are often based in countries without export credit facilities, and may be excluded from the bidders' list.
- The recipient country may be able to provide capital plant but does not qualify for export credit.
- The risk of a technical problem increases if suppliers are chosen on financial grounds alone.
- It may be necessary to group together several smaller purchases in order to qualify as high-value capital goods. This may reduce the number of available suppliers and increase the overall cost.
- The cost of reducing the foreign exchange risk can be high, as export credits are usually provided in US dollars or the currency of the exporter.
- The cost of the financial package can be increased by front-end fees, commitment commissions and guarantee premiums.

Figure 6.1. Export credit analysis performed by the project promoter

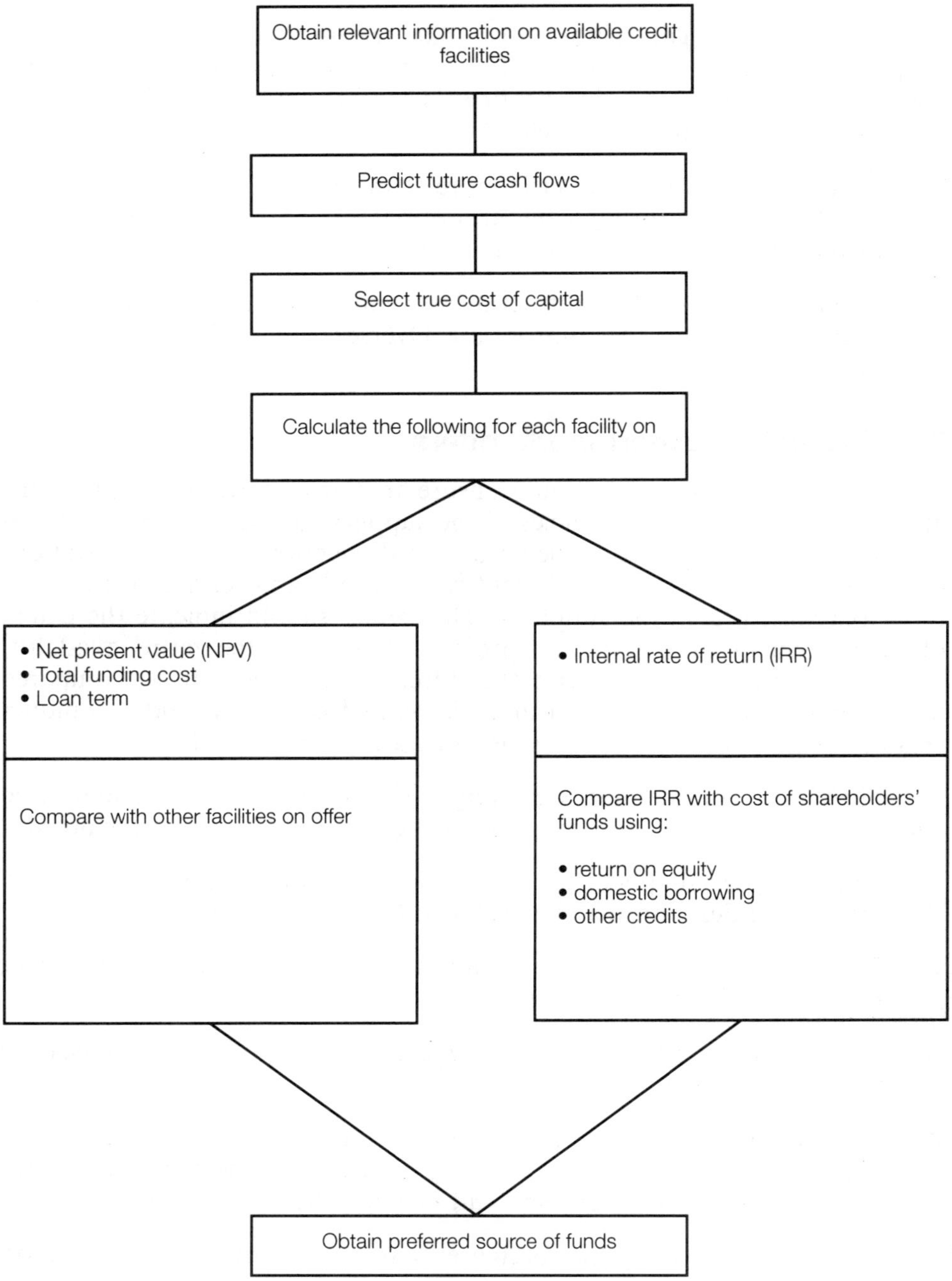

- It is difficult to predict the additional cost resulting from the high degree of coordination and cooperation required between the financial and the engineering and procurement divisions.

6.8 Summary

It is important to establish a set strategy to help define the project and select the appropriate funding approach. For example, the following strategy could be adopted:

- maximum utilization of export credit finance;
- top-up finance provided through Eurocurrency bank credit; and
- working capital obtained in local currency from a domestic bank.

Once offers have been received, it may be possible to obtain better terms by approaching export credit agencies which recognize various international conventions. For example, the Berne Union was established in 1934 and is an international association of credit insurers representing 26 countries. Its recent work has been directed towards avoiding a "credit race" between exporters from different countries by discouraging abnormally long credit or competition in awarding credits. Under the Berne Union's matching principle, each member is allowed to match terms offered by competing countries.

References

Bank of England, Overseas Department. 1979. *Money for exports* (London).

British Overseas Trade Board. 1991. *The Aid and Trade Provision: Guidelines for applicants* (London, Department of Trade and Industry).

Edwards, H. 1983. *Export credit: The effective and profitable management of export credit and finance* (London, Gower).

Einzig, P. 1973. *Roll-over credits: A system of adaptable interest rates* (London, Macmillan).

Gmur, C. J.; Budd, N. 1984. *Case histories of trade financing* (London, Euromoney).

Hesketh-Roberts, P. J. 1985. "Second Congress of Civil Engineers: The engineer in society. Finance of overseas contracts", in *Proceedings of the Institution of Civil Engineers* (London), Part 1, Vol. 78, June, pp. 580-582.

LeGault, R. 1989. "An exploration into overseas development", in *Real Estate Finance Journal* (Boston, Massachusetts), Vol. 5, No. 2, Fall, pp. 52-56.

Love, J. H.; Dunlop, S. 1990. "The domestic impact of overseas aid: A case study of aid and trade provision", in *National Westminster Bank Review* (London), Aug., pp. 54-68.

Munro, J. D.; McGrath, A. J. 1983. "Financing roadworks overseas: UK contractors' experience", in *Highway Engineer,* Vol. 30, No. 7, July, pp. 7-11.

Organisation for Economic Cooperation and Development (OECD). 1982. *The export credit financing systems in OECD member countries* (Paris).

Price, A. D. F. 1994. *International project accounting,* International Construction Management Series No. 1 (Geneva, ILO).

Royal Institute of International Affairs. 1981. *Export credit: The next phase,* Conference report (London).

Watson, A. 1983. *Finance of international trade* (London, Institute of Bankers, 3rd ed.).

Whiting, D. P. 1973. *Finance of foreign trade and foreign exchange* (London, MacDonald and Evans).

PART III

ALTERNATIVE APPROACHES TO FINANCING

JOINT VENTURE FINANCING

7

7.1 Introduction

The term "joint venture" is becoming more common in the construction industry, not only for overseas work but also in home markets. Very often joint ventures are seen as the best way to win contracts with minimum risk involved. There are times when joint venture tenders will appear more attractive to the client. However, if a joint venture agreement is entered into without sufficient consideration, the pre-contract stage could well highlight certain deficiencies, resulting in an unsuccessful tender.

In China there are over 9,000 cooperative ventures involving at least one foreign partner. A large proportion of these (65 per cent by value) involve either overseas Chinese (usually in Guandong and Fujian provinces) or Hong Kong investments. The United States and Japan are the next highest investors, followed by the European Union. Around 60 joint ventures involve British organizations.

This chapter reviews the main types of foreign investment and discusses the possible reasons for establishing a joint venture. The planning and selection processes are also described. Although most countries have regulations dealing with joint ventures, this chapter refers specifically to regulations relating to joint ventures in China. There are three main types of foreign investment:

- wholly owned foreign enterprises;
- contractual joint ventures; and
- equity joint ventures.

The latter two are discussed below.

7.2 Contractual joint ventures

Of the two main types of joint venture, the contractual joint venture is the more flexible one. In China, the amount of foreign investment can vary from a simple licence agreement for a manufacturing process to an arrangement which looks very similar to an equity joint venture with at least 25 per cent foreign investment. In most cases, the establishment of a contractual joint venture does not involve the creation of a new company separate from the contracting parties.

A hybrid joint venture shares many features of a contractual joint venture, but it is treated differently from a tax point of view. In a hybrid joint venture, the Chinese partner's equity often takes the form of labour and utilities provided.

7.3 Equity joint ventures

Equity joint ventures require a limited liability company to be established and registered, with equity investments from both a foreign partner and a Chinese partner. However, equity can take the form of equipment, or even technology and property rights. The foreign partner must invest at least 25 per cent, with the Chinese partner retaining a majority holding. In practice, most joint ventures are a 50/50 or 49/51 split between foreign and Chinese partners.

7.4 Staffing

Workers can be obtained by contacting local labour departments or recruited directly through the joint venture. When a trade union exists, staff have to be employed on a contract system between the joint venture and the union. When suitably qualified local staff are not available, foreign personnel can be employed.

7.5 When to establish a joint venture

Joint ventures are not the easiest type of contract to establish, manage and operate. Prior to adopting a joint venture approach, management should consider the more traditional subcontract approach and the advantages it provides. For example, in the subcontract documents, the liabilities, responsibilities and payment methods are well established and require less administration. However, a joint venture approach does have some good arguments supporting its application. For example:

- risks associated with very large projects are reduced;
- requirements for local participation are satisfied;
- all parties have equivalent status, which is not available through traditional subcontracts;
- the general resources of partners are combined;
- the specialist skills of partners are combined;
- the bonding requirements of partners can be shared;
- bid or prequalification credibility is increased;
- the commitment of resources from two or more companies is demonstrated;

- competition is reduced; and
- a transfer of technology can be achieved.

Most of the reasons for selecting a joint venture approach relate to either the size of the project or the associated risk. Obviously, there are also advantages associated with the inclusion of local parties and the transfer of technology.

At a very early stage, an organization should ask itself the following question before embarking on a joint venture:

- Could we prequalify for the project and successfully execute it by using subcontractors?
- If the answer is "no", then the reasons must be analysed. This will identify specific weaknesses which could be overcome by including the appropriate partner.

7.6 Participation

The first step for the various parties should be to decide the type of joint venture to be adopted. Once this has been achieved, the extent of individual participation can be established. This will involve the equity split based on inputs of capital, staff, plant and other resources, and does not necessarily have to be on an equal basis. The types of bond and proportions of profit also have to be agreed upon.

While agreeing the level of participation among the parties, a joint venture sponsor or leader has to be established. The partner with the larger equity share usually takes the leading role. However, this may not be the case when one partner has skills central to the success of the project or has a long-established relationship with the client.

It is essential that the sponsor's duties and powers of authority are clearly defined when the joint venture agreement is drawn up. There is always the possibility that the other parties may become suspicious and critical if the sponsor's administrative and managerial expenses are high. This is particularly the case if these costs are directly reimbursed and not paid as a fee. It is therefore sometimes beneficial if the burdens of contract administration are shared. When equal partners exist in a joint venture, it is probably better to appoint a joint venture management team controlled by the joint venture board rather than have a sponsor.

7.7 Prebid agreements

The process of tendering for a project can be both costly and time-consuming. It is thus important that some form of prebid agreement, which clearly states the terms of withdrawal, should be reached. It can be very expensive to establish a completed joint venture agreement and it is often a lengthy affair. However, some form of early agreement is necessary, as the client will look for a definite commitment with the tender documents (or even a prequalification) and the partners need to be confident that they are bound together. The prebid agreement needs only to be a brief document, but the following should be clearly stated:

- the partners;
- the level of participation;
- the sponsor or lead partner;
- the type of joint venture;
- extent of commitment to the project;
- the split of tender costs;
- the conditions of withdrawal;
- an exclusivity clause that binds the partners to the joint venture and prevents them from bidding for the work in the form of another joint venture or by themselves; and
- a specimen joint venture agreement may be appended to illustrate the anticipated final form of contract.

7.8 Management structure

Both types of joint venture (i.e. contractual and equity) require similar management structures, which must be formalized as soon as prequalification has been achieved. The need for efficient management during the tender period and beyond is essential.

Usually, the principal decision-making group takes the form of a Management Board or Committee comprising one member representing each of the partners. A general manager, appointed by the Board, would have the responsibility to the Board for the running of, and reporting on, the joint venture. The Chairpersons or Managing Directors of the various partners should form themselves into a small policy board in order to review overall progress and performance.

In the event of a dispute arising on the Management Board, a right of appeal to the policy board should be available. Most disputes that arise are not suitable for arbitration as the costs will influence the joint venture. However, the Management Board should include individuals of sufficient seniority to commit their respective companies to certain policy decisions without constantly referring back for approval.

Separate groups of specialist advisers should be available to both the Management Board and the General Manager when certain legal and financial issues have to be considered. Typically, a joint venture management structure will take the form illustrated in figure 7.1.

Figure 7.1. Joint venture management structure

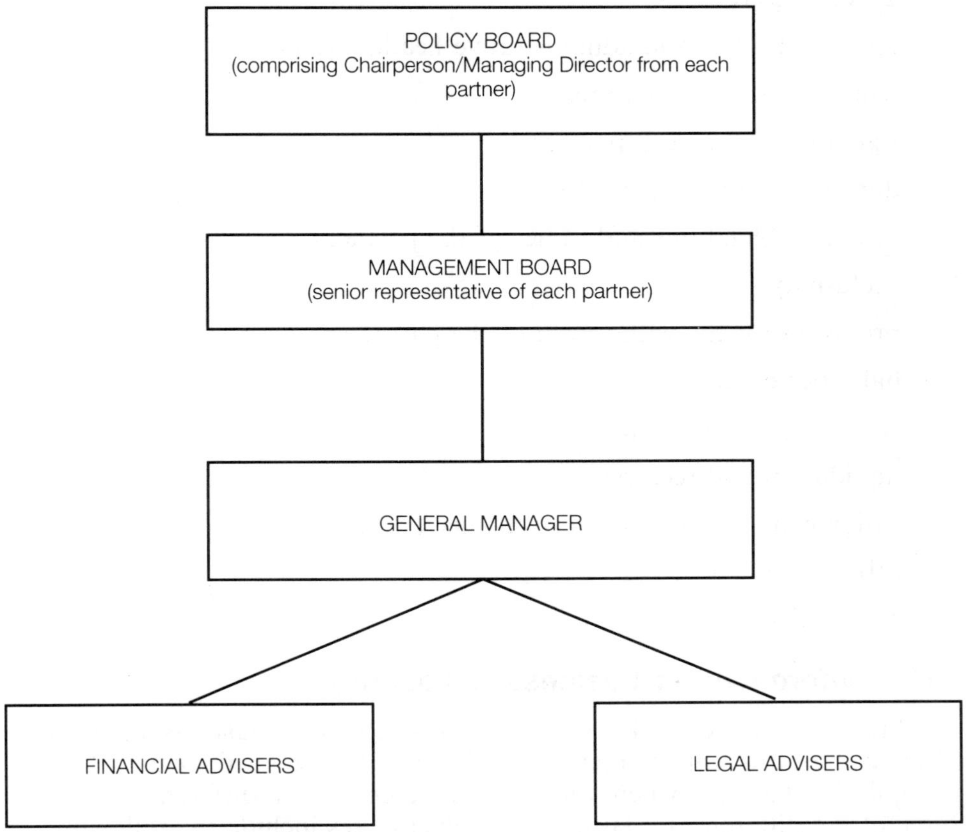

7.9 Joint venture documentation

It is important to realize that foreign law could impose a system upon the joint venture and override the joint venture agreement or documentation. It is up to the various partners to agree on which country's laws they wish to be governed by. This concerns specifically the partners' relationships with each other and does not necessarily have to relate to the country of the partners or the country of operation.

This book has been directed towards the financial aspects of international construction. The intention in this chapter has not been to provide the reader with detailed information relating to the different types of contractual arrangement and documentation. These are discussed at length in other books in this series. However, the information that one would ideally wish in joint venture documents is listed below:

- recitals;
- governing law;
- registered office and administrative headquarters;
- name of the joint venture;
- signatures from each partner;
- duration of the joint venture;
- powers, obligations and duties of the partners;
- exclusivity;
- provision for the withdrawal of the partners;
- liabilities of the partners;
- special liabilities of the sponsor;
- liquidation and receivership;
- assignment (i.e. allocation of finance and profit);
- arbitration; and
- registration.

7.10 Differences in business philosophy

Early on in the tender stage, the differences in business philosophy adopted by different companies will have to be dealt with. This is particularly the case when the partners come from different countries. Some of the differences relating to financial issues include the following:

- how capital equipment should be written down;
- cash-flow expectations;
- assessment of risk;
- conditions of contract;
- acceptable profit levels;
- allocation of overheads;
- type of labour; and
- management style.

7.11 Summary

The main advantage of a joint venture is that the agreement minimizes the risk to each of the parties involved. The sponsors' liabilities are restricted to the extent of their own investment and advances made to the project. However, they may sue each other if there is a breach of contract.

As common tenants, each party has an undivided interest in the assets of the joint venture.

Joint ventures may be used to avoid balance sheet exposure and, according to accounting standards, sponsors should limit their involvement to between 20 and 50 per cent. If this is the case, a single-line entry can be used to state the level of investment. When the sponsor's involvement is more than 50 per cent, a detailed presentation is required; consequently, the financial accounting procedures that apply to corporate ownership also apply to joint ventures. However, the Financial Accounting Standards Board in the United Kingdom may introduce partial consolidation (i.e. presentation of financial statements), thus increasing balance sheet exposure.

In most cases, sponsors arrange their own finances for their share of the joint venture. The sponsors may also establish a new corporation to act as an operating agent, but this does not usually constitute a legal entity that can easily borrow on its own account.

References

Armitt, J. A. 1985. "Joint ventures: Formation and operation", in *Management of international construction projects*, proceedings of a conference organized by the Institution of Civil Engineers and held in London on 14-15 November 1984 (London, Thomas Telford), pp. 61-71.

Beamish, P. W. 1992. *Multinational joint ventures in developing countries* (London, Routledge).

Beidleman, C. R., et al. 1990. "On allocating risk: The essence of project finance", in *Sloan Management Review* (Cambridge, Massachusetts), Spring, pp. 47-55.

Black, D. M. 1981. "To joint venture, or not – Is it for your firm?", in *Issues in Engineering* (New York, American Society of Civil Engineers), Vol. 107, No. 1, Jan., pp. 11-23.

Brown, R.; Spires, M. 1988. *Partnership schemes and joint ventures between the public and private sectors,* Stamford Seminar (London, Capital Planning).

Fordham Corporate Law Institute. 1990. *International mergers and joint ventures,* 17th annual proceedings of Fordham Corporate Law Institute (New York, Transnational Juris, Fordham University).

Gualtieri, G. 1991. *The impact of joint ventures on competition. The case of the petrochemical industry in the EEC*, Final report to the Commission of the European Communities. (Brussels, Office for Official Publications of the European Communities).

Ministry of Labour, China. 1984. *Provisions for the implementation of the Regulations on the Labour Management of Joint Ventures using Chinese and Foreign Investment* (Beijing, 19 Jan.).

Provisions of the State Council of the Peoples' Republic of China for the Encouragement of Foreign Investment. 1986. (Beijing).

Razvigorou, E.; Wolf-Laudon, G. 1991. *East-West joint ventures: The new business environment* (Oxford, Blackwell Finance).

Reed, C. 1988. "Greece bids huge irrigation complex", in *Engineering News Record,* Vol. 221, 18 Aug., pp. 59-60.

Sinclair, K. 1990. "China projects: Daya Bay nuclear plant on course", in *Asian Finance* (Hong Kong), Vol. 16, No. 3, 15 Mar., pp. 20-21.

Watkins-Mathys, L. 1992. *The role of joint ventures in East-West trade up to 1990,* PhD thesis (Loughborough, Leicestershire, Loughborough University of Technology).

BARTER ARRANGEMENTS

8

8.1 Countertrade

Countertrade usually refers to reciprocal and compensatory trade agreements involving the purchase of goods or services by the seller from the buyer of his or her products. Countertrade is thus a sophisticated marketing and financing device which can be used to:

- maintain a balance of payments;
- gain access to new markets;
- facilitate foreign investment; and
- transfer technology.

Countertrade encompasses all forms of reciprocal trade, some of which are listed below and described in this chapter:

- barter;
- counterpurchase;
- compensation;
- switch trading;
- offset; and
- bilateral clearing.

Most countertrade arrangements have the following characteristics:

- they are based on a single contract between the direct parties;
- they are completed in a very short period of time in order to reduce the impact of future price fluctuations; and
- they are made on a government-to-government basis through bilateral clearing agreements.

The findings of a recent United States International Trade Commission (USITC) study of countertrade demonstrated that "military countertrade is overwhelmingly directed towards NATO countries as well as Spain,

Australia, South Korea (Republic of Korea), Switzerland and Israel" and that "on the civilian side, countries from China to Finland and Austria to Canada make the top ten. Australia, Canada, China and Israel are the big four in terms of sales while Canada, Mexico, Australia and Brazil are the four largest in terms of obligations".

8.2 Basic barter arrangements

A basic barter agreement is the exchange of goods or services between two separate parties, usually countries or multinational firms, without the use of currency. This is, however, a narrow definition, as money usually comes into the deal at some stage.

Barter is a very broad area whereby any type of goods or services can be exchanged for one another. As long as someone wants what you have to offer, then an agreement can be reached. The exchange process should be completed within a short period of time, in order to minimize the impact of world price fluctuations. Longer-term deals tend to be risky because allowances have to be made for changes in market demand and supply.

Consider the following as an example of pure barter. The price of a barrel of a certain chemical is US$100, while the price of plaster is US$8 per bag. If the plaster manufacturer requires the chemical for some reason, then it may arrange a barter with the chemical manufacturer (so long as the chemical dealer wants plaster). Thus, if the plaster manufacturer offers 500 bags of plaster at an exchange rate of 1:12.5, then it would be entitled to 40 barrels of the required chemicals.

Once the amount of barrels and bags has been firmly agreed and delivery details fixed, a "swap" can take place. Although no money changes hands, the transaction does have a monetary value which is needed for the purposes of taxation, accounts, value added tax (VAT), and import/export duties. Currently, most transactions are done through money, either in its real or electronic form; hence there appears to be little place for true barter. However, there are several different types of barter, each with its own unique characteristics. The most common ones are discussed below.

8.3 Counterpurchase

Counterpurchase is a common type of countertrade favoured by developing countries, and is often used to safeguard their balance of payments. It involves an agreement between an importer and an exporter. The initial exporter will either buy the goods or services on offer, or will guarantee to find a buyer for the initial importer of its own goods. Deals involving only goods are called pure barter. Counterpurchase periods generally last between one and five years, with the agreement made as part of the original contract or as a separate but parallel contract.

8.4 Compensation

Compensation is an agreement, usually between the exporter and the importer, whereby the initial exporter accepts output from the exported product as full or partial payment for the goods initially provided. This method can be used to facilitate the sale of large turnkey factories and is becoming more popular. The average value of compensation arrangements tends to be considerably higher than for other types of countertrade (excluding offsets). The take-back periods tend to vary between five and 20 years. If the repayment occurs in the form of related goods, then it is referred to as "buyback".

8.5 Switch trading

Switch trading is used by some countries in order to liquidate trade imbalances. It is a process by which one party to a countertrade agreement can sell off purchase options (for goods or services) at a discounted price to a third party, who must complete the arrangement for the sale of goods. Some companies specialize in switch trading; the discount could be as much as 40 per cent depending on the demand for the goods. Vienna is the home of most of the main switch trading houses.

8.6 Offset

Many offset arrangements have been associated with the sale of foreign military items such as guided missile systems. Offset arrangements usually involve multinational enterprises from one of the industrialized countries and a foreign government. Obviously, most arrangements will have to obtain the home government's agreement in order to export certain high-tech information. Most offset arrangements exhibit a combination of characteristics associated with both counterpurchase and compensation. There are many options available, including:

- technology transfer;
- co-production;
- licensed production; and
- subcontract production.

8.6.1 Technology transfer

The transfer of technology often involves a countertrade or offset agreement. It normally provides technical assistance and training, research and development or patent agreements. Most developing countries have to make use of this approach if they are to bring their technology in line with that of the more advanced countries.

8.6.2 Co-production, licensed production and subcontract production

These types of barter are again on an international scale between either governments or producers. They usually enable one party to acquire technical information in order to manufacture certain complex products. There are many variations that can involve joint ventures or subsidiary arrangements enabling a foreign producer to manufacture a product without the actual transfer of important technology.

8.7 Bilateral clearing

The purpose of bilateral clearing is to help control international trade arrangements between countries. The respective governments make trade agreements in order to avoid transferring foreign exchange reserves between one another. If there is a trade imbalance, the accounts are settled annually. The balance of payments and exchange rates can also be controlled by using this type of barter.

8.8 Construction barter arrangements

Examples of construction barter arrangements between two countries are given below:

- In 1979, China agreed a barter with Australia for the construction of six 200-room prefabricated hotels. In return, China offered frozen and canned food.
- In 1981, Turkey arranged a barter with the United States for the building of two large plants for the production of PVC rolls. In return, Turkey provided artificial leather.
- In 1983, China bartered with Japan for the construction of oil pipelines. In return, Japan received crude oil and refined products.

Pure barter arrangements include only the exchange of goods or services, but these are definitely in the minority today. Although the definition of barter states that money is not involved or exchanged, this is not always the case and most agreements are far more complicated than simple barter.

There often is a time lag between the initial sale of goods and the final purchase of the exchanged product. In order to accommodate large delays, two different approaches can be taken. The first involves no exchange of money, but inflation is taken into account by adjusting the ratio of goods exchanged. Alternatively, the goods will be purchased as they are supplied, but the prices have to be agreed prior to the initial purchase. Thus, countertrade agreements are often part barter and part cash. The following example, illustrated in figure 8.1, demonstrates the involvement of cash in a barter deal.

Figure 8.1. Cash involvement in barter deals

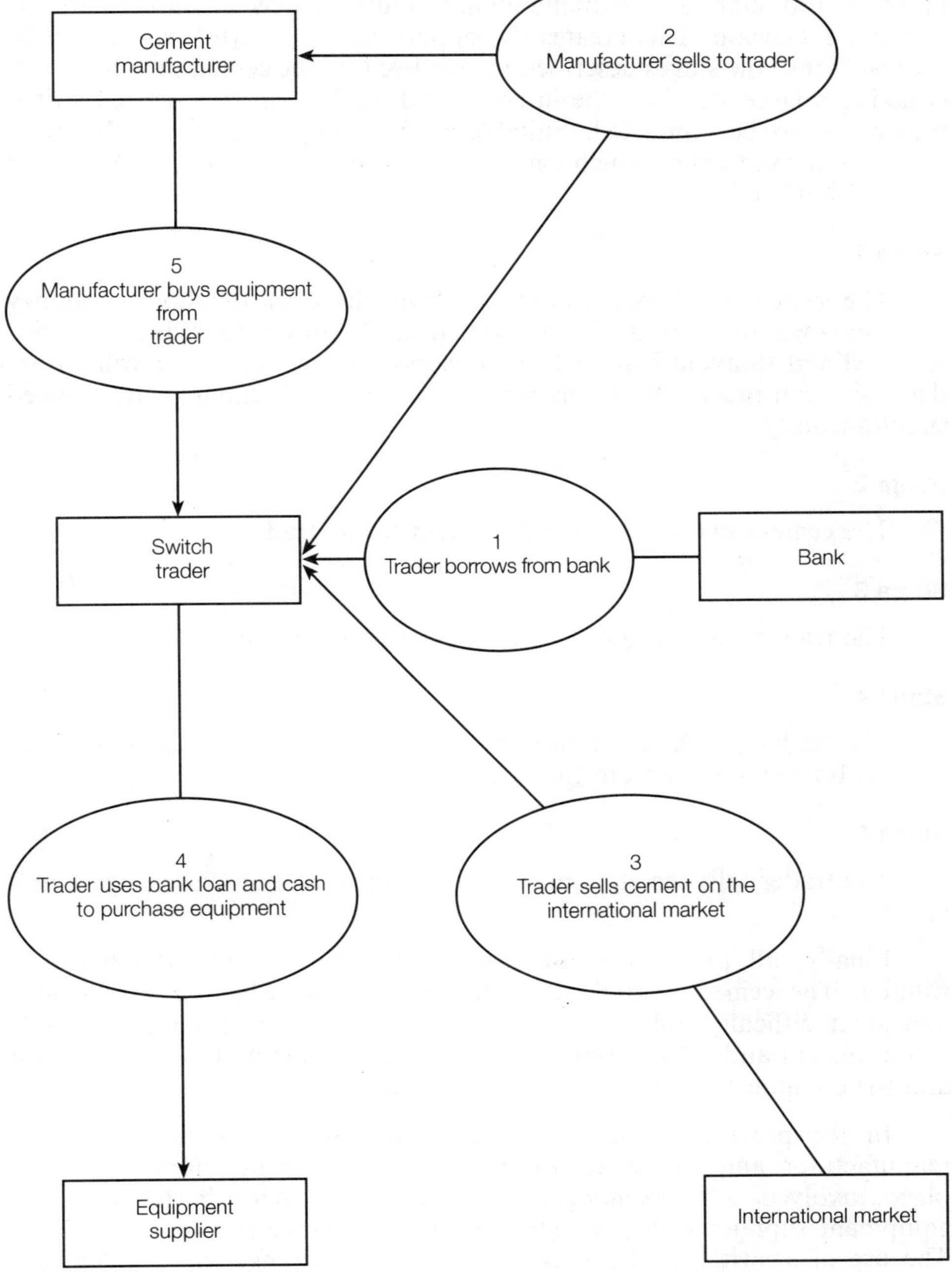

A small cement manufacturer wishes to expand into the international market. Although more money can be made on the international market, this is somewhat offset by the increased risks. In addition to selling cement in the world market, the manufacturer requires new equipment for its planned expansion. This creates the opportunity for barter, which can be separated into the stages described below. Because the cement manufacturer is taking a large step into the international market, an experienced switch trader has to be contacted. Suitable traders may well have substantial capital and fixed assets which can be used to reduce the risk involved and locate suitable clients.

Stage 1

The switch trader borrows money from the bank or allocates money from its own reserves to finance the deal. In order to reduce the risk involved and to avoid being left with excess cement, the trader will ensure that all contracts for exchange of goods and money are signed simultaneously.

Stage 2

The cement manufacturer sells cement to the trader.

Stage 3

The trader sells the cement on the international market at a profit.

Stage 4

The trader purchases equipment from a supplier and may well again have to borrow money from the bank.

Stage 5

The trader sells the equipment to the cement manufacturer, again at a profit.

Finally, all the parties are satisfied and have obtained what they wanted. The cement manufacturer has equipment, which it would have had great difficulty obtaining. The trader has made profits from the sale of the cement and the equipment. The bank has made money from interest and the equipment supplier has made an extra sale.

In the previous example the barter deal was between the cement manufacturer and the switch trader. The other transactions that took place involved cash exchanges. However, in a pure barter deal the equipment supplier would accept cement in exchange for goods supplied. The use of a switch trader may be required to arrange the deal but not necessarily to finance the process, as the money involved has just gone around in circles. Assuming that everyone uses the same bank, then the flow of the money can be summarized in the following six steps, as illustrated

Figure 8.2 Money flow in barter deals

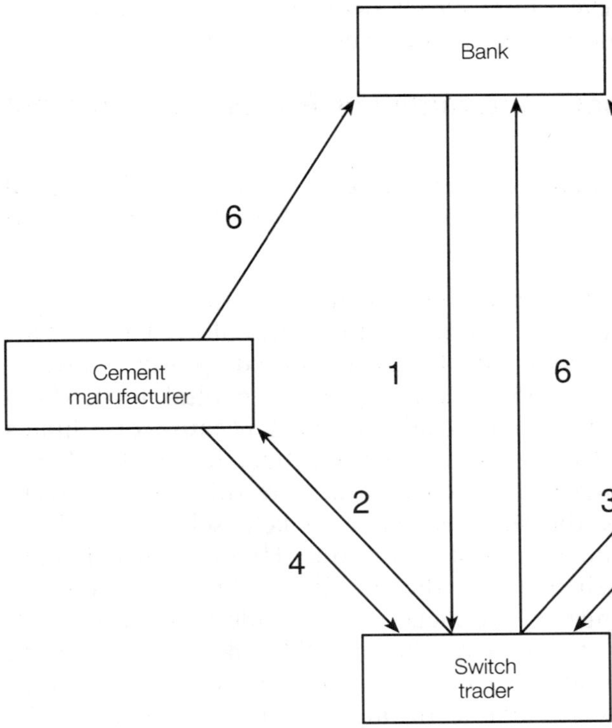

in figure 8.2. In this example, only the trader was required to borrow money from the bank, and even this could have been avoided if the trader had had sufficient funds or all exchanges took place simultaneously.

1. The switch trader borrows money in advance.
2. The switch trader buys cement, and some money goes to the manufacturer.
3. The switch trader buys equipment, and some money goes to the equipment supplier.
4. The cement manufacturer buys equipment from the switch trader, and some money goes to the switch trader.
5. The equipment supplier buys cement from the switch trader, and some money goes to the trader.
6. All profits go back into the bank.

In the pure barter deal the value of cement and equipment would be equal, and the exchange of goods performed simultaneously, thus removing the flow of any cash. However, pure barter deals are hard to achieve, so the process illustrated in figure 8.1 is the more common approach taken.

8.9 Case study: Huaneng International Power Development Corporation

Examples of barter transactions are easy to come by for general exchanges of goods and services, but they are significantly harder to find for construction projects.

Huaneng International Power Development Corporation, a Chinese company, has made extensive use of countertrade, particularly barter. This large organization imports and constructs coal-fired power stations. Huaneng was established in 1985 with a registered capital of US$100 million and ventured into the countertrade business because of China's foreign exchange shortage. Countertrade now accounts for 35 to 60 per cent of each contract. The terms of each contract are different for every power station, but basically the suppliers have to help sell or buy back Chinese-made products within an agreed time frame. Huaneng usually pays for its power stations in cash first and the suppliers fulfil their buy-back obligation later. A wide range of products are available to choose from, which makes the arrangement attractive. During 1986 and 1987, Huaneng imported just over US$2 billion worth of coal-fired power plants. Cheap foreign finance has only been available through the use of countertrade, and this has shown the way forward for financial reforms in other sectors (do Rosario, 1987).

8.10 Summary

Countertrade arrangements are not very appropriate as a means of financing construction projects, for the reasons listed below:

- only very large organizations can exploit countertrade to its full extent;
- countertrade arrangements are extremely difficult to arrange and can take a long time to finalize, time scales of a year or so being not uncommon; and
- in construction, definite starting dates are needed and a work schedule has to be produced. However, barter places so much responsibility on outside organizations that establishing a deal is just too time-consuming.

In most cases, it is best to avoid countertrade unless there are no conventional financing methods available, or there are real market advantages that occur from the deal. However, although there are currently only a few examples of construction projects funded through barter deals, the trend is increasing. Countertrade is here to stay and there is no choice but to become involved and learn how to operate effectively in this type of environment.

References

Alexandrides, C. G.; Bowers, B. L. 1985. "Finding markets for countertrade products: How to make trade statistics work for you", in *Countertrade and Barter Quarterly* (New York), Vol. 5, Spring, pp. 51-53.

——; ——. 1986-87. "Trade and countertrade in China", in *Countertrade and Barter Quarterly,* Vol. 12, Winter, pp. 41-44.

Countertrade Outlook. 1986. *Directory of organizations providing countertrade services* (Alexandria, Virginia).

do Rosario, L. 1987. "A new power centre: China's Huaneng cuts through roadblocks to boost energy output", in *Far Eastern Economic Review* (Hong Kong), Vol. 138, 17 Dec., pp. 116-118.

Hammond, G. T. 1990. *Countertrade, offsets and barter in the international political economy* (London, Pinter Publishers).

Healey, N. 1990. "Back to bartering basics", in *Accountancy* (London), Vol. 105, March, pp. 183-184.

Newman, S. G. 1985. "Co-production, barter and countertrade: Offsets in the international arms market", in *Orbis* (New York), Vol. 29, Spring, pp. 183-213.

Rubin, S. M. 1986. *The business manager's guide to barter, offset and countertrade, The Economist Intelligence Unit, Special Report No. 243* (London, Economist Publications Ltd.).

Siegel, J. G.; Wolitzer, P. 1986. "The accountant's role in barter arrangements", in *National Public Accountant* (Alexandria, Virginia), Vol. 31, No. 8, Aug., pp. 34-37.

Stern, P. 1985. *US International Trade Commission Report,* No. 332, 185 (Washington, DC, USITC).

Verzariu, P. 1985. *Countertrade, barter and offsets: New strategies for profit in international trade* (New York, McGraw-Hill).

BUILD-OPERATE-TRANSFER OPERATIONS

9

9.1 Introduction

Build-Operate-Transfer (BOT) as a means of project financing is a relatively new concept and falls under the general heading of concession financing. BOT involves the creation of a company not only to handle the construction of a project, but also to operate the new facility for a predefined time period. The arrival of BOT projects stems from governments wishing to avoid the risks associated with the financing of major construction projects. The BOT model is often selected because it prevents losses and reduces the danger of accumulating long-term debts. BOT has also been promoted as a way for developing countries to build infrastructure projects without having to finance the scheme from public funds.

BOT often involves the temporary privatization of public construction projects. The scheme is built and operated by a private developer, and finally transferred back to the government after a substantial time period. The developer is given a certain number of years of positive revenue to pay back any investment before the government takes over. The normal time period is around 20 years. However, Eurotunnel has been granted a 55-year concession on the Channel Tunnel project before it reverts back to the British and French Governments. Build-Own-Operate (BOO) projects do not revert back to the government.

9.2 Establishing project viability

BOT operates differently throughout the world, but there are similar characteristics when it comes to creating a financial package. First of all, the viability and profitability of the project has to be established. Second, the project has to be sold to the international money market before work can start. As some previous BOT projects have been unsuccessful, most investors regard BOT projects as very risky and have to be persuaded that the project is a sound investment. The important steps that are required to safeguard investors' money are listed below:

- identify the project and justify the need for it;

- estimate the project cost;
- estimate the project income;
- determine if the project is economically viable;
- agree upon the concession period;
- agree upon safeguards (i.e. the government could make up revenue shortfalls);
- establish a project company;
- arrange a debt package with a bank, possibly in the form of a syndicated Eurocurrency credit or a syndicated letter of credit facility;
- establish an underwriting group for local and international equity investors;
- invite construction contractors to hold equity in the project; and
- sell equity in the project as required during the construction period.

Common to all types of BOT project is the element of equity put into a special-purpose vehicle (usually a consortium including the contractor, the operator and the buyer of the end product) by the project sponsors. The equity achieves three purposes. First, it acts as an incentive for the contractor to complete the project on time and within budget. Second, it demonstrates the sponsoring consortium's faith in the economic viability of the project, thus encouraging lenders to provide additional finance. Third, it acts as a cushion in the event of the project proving to be less economic than expected.

9.3 Associated risks

BOT is a new approach to project finance. It appears very attractive, as contractors can generate their own work and investors can make substantial profits. However, as with any innovation, there are also associated risks, and the extent of these must be established before investing in such schemes. The types of risk taken by the sponsors and others associated with BOT projects are presented below:

- *Reservoir risk*

  Will there be enough raw materials for the construction? By their very nature, BOT projects tend to be large and have extensive labour and material requirements. It is important that local sources can satisfy the demand. If material or labour has to be imported, then the project cost will escalate rapidly.

- *Market price risk*

  It is important that the toll prices (i.e. the charge for use of a service) are neither too high nor too low. Tolls set too high may generate insufficient volumes (e.g. traffic flow) to make a profit; tolls set too low may create a high volume, but not enough to remove the debt and give a decent return on equity. Predicting the correct market price is one of the major risks associated with most BOT projects.

- *Completion risk*

  Although this applies to most construction projects, BOT projects have an extra need to be completed on time and within budget, as no income can be generated prior to completion.

- *Operating risk*

  The predictions of running and maintenance costs need to be accurate, and revenue generated should exceed them. This is not easy, as most BOT projects are unique and have substantial life spans.

9.4 Case study: The Channel Tunnel

One of the best case study examples of a BOT arrangement in today's economic climate is the Channel Tunnel. The tunnel is being funded on a limited recourse basis, with contractors being paid and the debt serviced by the cash flow generated by the completed project. The project is an entirely private sector affair. The British and French Governments have both stated that they will not bear the financial risks should the project run into difficulty. The Channel Tunnel Group, Eurotunnel, is a private company created for the project and has been granted a 55-year concession on the tunnel traffic before the two governments take over. However, this was only awarded after the winning consortium took substantial equity in the project, thus demonstrating its faith in its economic viability. This equity also creates a big incentive to complete the project on time and within budget.

The Eurotunnel venture is extremely risky because of the very nature of the project. Normally, if something goes wrong the project or its assets can be sold off to recoup some of the investment. However, if the tunnel fails, it will be difficult to sell a hole between two countries.

The cost of financing the project was initially estimated to be over £6 billion, making it the largest venture ever. £5 billion of this was in the form of a syndicated loan provided by 40 banks, led by NatWest in the United Kingdom and Crédit Lyonnais in France. The loan was issued in three tranches, subject to the raising of £1 billion equity. The first tranche of £46 million was provided by the 14 original project promoters and the main lending banks. The second tranche of £206 million was raised in October 1986, with £70 million coming from British and French markets. The remaining £770 million was obtained towards the end of 1987 through the public share offer. The 1990 rights issue raised £566 million. The total equity now stands at £1.588 billion.

Drawdowns were agreed in three- and six-month advances at the following rates:

- the first £4 billion was priced at LIBOR + 1.5 per cent, falling to 1.25 per cent on completion;
- the next £2.3 billion was priced at LIBOR + 1.75 per cent, falling to 1.5 per cent on completion;
- the final £500 million is currently priced at LIBOR + 2.5 per cent, falling to 2.25 per cent on completion.

Eurotunnel has been financed mostly in French francs and sterling, as the eventual revenues will come from these currencies. This approach applies to most projects. Petrochemical plants are normally financed in dollars; and major hotels may be financed using a combination of different currencies, as its residents will be of different nationalities. The Eurotunnel project is being watched closely by people all over the world, and the reputation of BOT as a method of project finance is at stake.

9.5 Case study: The Malaysian North-South Highway

The 900-km North-South Highway project in Malaysia is another good example of how BOT can be used to finance projects that are viable and profitable but cannot be financed by the national government. It was initially intended that the scheme would be built by public sector construction, but this fell through after the Government realized that it could no longer afford to finance the project, and BOT was almost the only method of getting the job finished. In 1987, United Engineers Malaysia (UEM) awarded a concession contract worth just over US$2 billion.

Under the concession, roads already built and owned by the Malaysian Highway Authority were taken over by the new operating company. Part of the deal was that modernization and upgrading, such as improved toll booths and lighting, were carried out on existing roads. UEM attached conditions of its own to the deal, the main one being that if traffic flows fall below an agreed level, the Government will reimburse the lost revenue. The project should take seven years to complete, starting in 1987, and any extra financing that is required will come from equity.

Once the concession contract was secured, UEM created a subsidiary company, Project Lebuhraya Utara-Seletum (PLUS), to act as concession holder and borrower for the project. A 30-year concession period was granted, with anticipated net dividend earnings of over US$400 million. In addition to this, UEM should make over US$150 million pre-tax profits from the project during its construction period. (A particular advantage of BOT road projects is that income can be generated during the construction phase, which is not possible on bridge and tunnel projects.) Some support has come from the Government in the form of a US$60 million loan drawn down over 11 years. The interest rate is 8 per cent per year and the loan has a 15-year grace period.

9.6 Case study: The Sydney Harbour Tunnel

The Sydney Harbour Tunnel was initiated to take the strain of the cross-harbour rush hour traffic which the Harbour Bridge could no longer cope with. Westpac provided a finance deal which took two years to put together. The tunnel was built and is being operated by Transfield Ply Ltd. and Kumagai Gumi, but ownership will revert back to the New South Wales Government in the year 2022 when the project will be debt free and will have an estimated value of A$2 billion. Most of the project finance came from the following sources:

- A$501 million worth of 30-year bonds, which were issued in eight tranches over a five-year period. The bonds were placed privately by Ord Westpac's fixed interest brokerage arm;
- A$17 million paid by the joint venture partners for the first year's construction, and an A$223 million loan made by the state government to cover most of the second-year costs;
- an A$40 million performance bond against cost overruns and construction delays to the project was lodged by the joint venture partners.

The project revenue will come mainly from the tunnel tolls which the operators have agreed to limit to A$1 (at 1986 prices) throughout the project life. Increased bridge tolls will also be used to generate project revenue. Because the toll prices have been fixed, the Government has underwritten minimum traffic flows for both the bridge and the tunnel. However, in order to prevent Transfield Ply Ltd. and Kumagai Gumi enjoying excessive profits, clawback provisions have been included.

Fund-raising costs were substantially reduced because public equity insurance and loan syndications were both avoided, unlike the Hong Kong Tunnel and Channel Tunnel projects. The following features of the project's financial package created a favourable reaction to the 30-year securities:

- the extended maturity, longer than anything previously available in the Australian market;
- an undisclosed yield, believed to be close to 6 per cent, with interest payments protected by an index-linking formula;
- repayments of principal with quarterly interest instalments, previously available in Australia only through the fledgling domestic mortgage-backed securities market; and

- the eight-tranche investment schedule, enabling institutions to lock in future paper pick-up (i.e. investors' agreement to provide funds) matching cash-flow expectations.

9.7 Summary

The concept of BOT has been around for a long time, but only recently has its full potential been realized. Over the past decade have been an increasing number of projects constructed using BOT as a method of financing. During the next decade, the greatest potential for BOT will be in developing countries, where there is the need for large infrastructure projects but limited government money to finance them.

References

Barrett, M. 1986. "Project finance develops new risks", in *Euromoney* (London), Oct., pp. 73-85.

——. 1987. "Putting your equity on the line", in *Euromoney*, Oct., pp. 119-128.

Carnevale, F. 1989. "Putting BOT on the road", in *Euromoney*, Apr., pp. 87-90.

Goltz, T. 1990. "Program with 22 dams suffering", in *Engineering News Record*, Vol. 224, 1 Feb., pp. 40-41.

Ingersoll, R.; Goltz, T. 1990. "Industry heading East", in *Engineering News Record*, 1 Feb., pp. 27-28.

Lowenstein, J. 1988. "Revival in Aussie France fortunes", in *Euromoney, Special Supplement*, pp. 43-48.

Nisse, J. 1988. "Survival of the strongest", in *The Banker* (London), Jan., pp. 71-72.

"Philippine BOT law update: Implementing rules and regulations for foreign investments act of 1991". 1992. *East Asian Executive Reports* (Washington, DC), Vol. 14, No. 2, 15 Feb., pp. 8, 13, 22-27.

Reina, P. 1992. "Bangkok builds BOT megaprojects", in *Engineering News Record*, Vol. 228, No. 24, 15 June, pp. 34-36.

Sington, P. 1989. "Project finance: Limited recourse shifts the risk", in *Euromoney*, Apr., pp. 82-88.

Stone, E.; Cross, M. 1990. "Construction: A revolution in the building; Japan keeps home market to itself", in *Asian Business* (Hong Kong), Vol. 26, No. 12, Dec., pp. 58-69.

Tavakoli, A.; Tulumen, S. C. 1990. "Construction industry in Turkey", in *Construction Management and Economics* (London), Vol. 8, No. 1, Spring, pp. 77-87.

Tiong, R. L. K. 1990a. "Comparative study of BOT projects", in *Journal of Management in Engineering* (New York), Vol. 6, No. 1, Jan., pp. 107-122.

——. 1990b. "BOT projects: Risk and securities", in *Construction Management and Economics*, Vol. 8, No. 3, Autumn, pp. 315-328.

"Turkey – The economy: Lower project spending, price increases, higher growth than forecast; Turkey – Investment and BOT projects". 1988. *Middle East Executive Reports* (Washington, DC), Vol. 11, No. 6, June, pp. 19-22.

Williamson, J. 1988. "Turkey heads into the next decade with a 21st century telecom system", in *Telephony* (Chicago), Vol. 215, No. 26, 26 Dec., pp. 20-28.

INSURANCE AND PERFORMANCE BONDS 10

10.1 Introduction

This chapter discusses some of the many different types of insurance and bonding that are either available to the construction industry or are required by law. A basic guide to insurance is given in order to familiarize the reader with the insurance issue. A description of the constructor's legal obligations and a brief outline of how these obligations may be covered using insurance are also included. These entail matters such as public liability and accident and injury to workers, as well as personal indemnity. After each description of the legal aspects, suggestions are given as to the type of insurance policy which would be required in order to cover these liabilities and, where possible, an estimation as to the size of cover. The role of performance bonds, which are similar to insurance, as well as the different types of bond available, are also dealt with towards the end of this chapter.

10.2 Insurance policy

An insurance policy is a form of guarantee or coverage taken out by either an individual or a company in order to cover financial loss. There are many factors that can generate a financial loss, for example, the insured not being able to fulfil a contractual undertaking, or legal action being taken against a company or an individual on account of public liability or damage to property. One of the most widely used insurances is that of property against such contingencies as fire, vandalism or theft. If a company does not appreciate the extent of its liability and fails to obtain the appropriate cover when undertaking a large construction project, it could well find itself facing a major financial loss, or even bankruptcy, as a result of legal action taken by a third party.

Insurance will always be an important consideration in the construction industry because of the high risk factor. There are many different types of policy available and construction policy wordings differ greatly from one another. The difference may be required to satisfy different intentions, but may also be the result of the same intention being expressed in a different way. The insured and their advisers may have their own requirements and certain policies will be drafted for specific projects. In the construction industry an insurance policy may be taken out because:

- it is compulsory by law, for example, motor insurance, employers' liability insurance or public liability insurance;
- it is required as part of a contract; or
- it is just common sense to do so.

An important characteristic of any insurance policy is the extent of cover available. In any policy there will be a list of exclusions indicating events that are not covered. These exclusions may include events for which it would prove too costly to provide cover, such as the outbreak of war. Events covered by other policies should also be excluded in order to keep premiums as low as possible.

10.3 Disclosure of material facts

Before an insurance policy is offered by the insurers, the insured is often required to complete a proposal form which asks a number of questions concerning the item to be insured. The person filling out the form must answer the questions to the best of his or her ability, and no information must be deliberately withheld. It is the legal duty of the insured to disclose any information which will affect the premium, whether or not the question is asked on the proposal form. By not disclosing these facts there is a good chance that if a claim is made on the insurance, and certain considerations come to light, then the insurer could consider the contract void. Consequently, the insurer is under no legal obligation to pay the claim. Set out below are general material facts that must be disclosed, if relevant, when attempting to insure a construction project:

- that the insured item is open to a higher risk than would normally be expected. An example of this could be if a building were to be constructed on land that is known to be susceptible to flooding;
- that the liability of the insurer is higher than would normally be expected. This could be appropriate to a contractor working on a building containing valuable items, at a location known to have a high incidence of theft;
- that the person applying for the insurance is not a fit character to receive it. This may be due to the person having a criminal record for offences such as arson or theft; and
- any other item which the insurer might consider material. This could encompass such things as a particular building site known to have a high number of thefts.

10.4 Insurance and liability

There is a wide variety of insurance policies available to the contractor. These may be classified as cover for liability, material damage, pecuniary, benefit, composite and combined, and miscellaneous. The most important policies cover the insurer's legal liability to third parties, and most of these are required by law.

Before looking at the main liability policies, it is essential to consider how the law sees your role in terms of a contractor or an individual. The very basis of common law is that if you commit a civil wrong against a third party, then you are held liable to that party and he or she is entitled to seek some kind of compensation from you through the law courts. In most countries, there are well-defined areas of liability where there is a right to compensation for injury, death or damage to property. The laws on liability are long and complex, but listed below are some of the situations in which liability arises.

10.4.1 Negligence

Negligence is the cause of most claims. Negligence is not doing something that a reasonable and prudent person would do, or doing something that a reasonable and prudent person would not do. It is important that the plaintiff establish a link between the act of negligence and the resulting accident. However, where an extraordinary event occurs, the burden of proof is passed on to the defendant. For example, if a builder constructed a wall, and in doing so built it so badly that the next day it fell on an innocent bystander, the builder could be sued under the liability law because, in the natural order of things, walls should not fall down.

If a claim of negligence is to be successful, the defendant must also owe a duty of care to the plaintiff, there must have been a breach of that duty, and the plaintiff must have sustained injury or damage as a result.

10.4.2 Nuisance

Nuisance can be defined as a wrong done to a person by unlawfully disturbing the person in the enjoyment of his or her property (i.e. a private nuisance) or, in some cases, in the exercise of a common right (a public nuisance). A nuisance can take the form of excessive noise, smell, vibration or dust. A public nuisance must involve a substantial number of people who are adversely affected and is a criminal offence. A private nuisance can involve an individual, but the period of time over which the nuisance occurs must be of a substantial length. In some cases, there may be an overlap between nuisance and negligence, as illustrated by the following points:

- both public and private nuisance involve an element of unlawful interference;

- it is generally accepted that private nuisance cannot result solely from personal injury. The injury must be caused in connection with the enjoyment of land;
- unlike negligence, a nuisance could be caused even if reasonable care has been taken;
- nuisance must involve some actual (not prospective) damage; and
- certain actions may be the result of nuisance caused by one party (the landowner) and negligence on the part of another party (the contractor).

10.4.3 Trespass

Trespass is an unlawful act committed with force and violence on the person or property of another. There are three types of trespass; trespass to land; trespass to goods; and trespass to persons. Indirect injuries are caused by acts of negligence or nuisance. Trespass must involve some form of "direct" injury.

10.4.4 Strict liability

Certain strict liabilities have been imposed by law, even though a defendant may not have intended to injure anyone and may have exercised reasonable care. Lord Cranworth summarized strict liability to be "when one person in managing his or her own affairs causes, however innocently, damage to another, it is obviously only just that he or she should be the party to suffer". This part of the law is best illustrated by the court case of Rylands v. Fletcher. In this case, a landowner employed a contractor to build a reservoir on his land. When the reservoir was finally constructed, the water broke through some old mines and flooded a neighbour's property. The neighbour sued and the landowner was held responsible because the court ruled that "any person who for his own purposes brings on his or her lands and collects and keeps there anything likely to do mischief if it escapes must keep it at his or her peril and if he or she does not do so is *prima facie* answerable for all the damage which is the natural consequence of its escape".

10.4.5 Breach of statutory duty

In construction especially, there are a number of rules laid down in law that govern how a company can act, the purpose of which are to prevent death or injury to the employees. British regulations include the Construction Regulations 1961, the Health and Safety at Work Act 1974, the Fatal Accidents Act and the Factories Act 1961. The main difference between negligence and breach of statutory duty lies in the degree of care required. The required standard of care is decided by the courts where the appropriate statute defines the standard of care required. Any death or injury caused by a breach of these laws could entitle employees or their families to sue for liability.

10.4.6 Vicarious liability

A contractor will have direct control of his or her employees and is therefore vicariously liable for their acts or omissions during their employment. Generally, the contractor will not have this degree of control over independent contractors and is not liable for their actions. However, there will be exceptions provided that the liability arises from the performance of the contract.

10.4.7 Product liability

In 1988, the EEC Directive on Product Liability came into force, whereby producers of goods can be held liable if the goods are defective and cause personal injury or damage to personal property.

10.4.8 Liability under contract

The previous torts are liabilities imposed upon the parties by law. Contractors can increase their legal liability by entering into a contract such as the Institution of Civil Engineers (ICE) Conditions of Contract, whereby the contractor agrees to indemnify the employer for damage to the works or in respect to third party claims. If someone is injured or his or her property damaged during an engineering project, and sues for compensation, the defendant often attempts to reduce the blame by bringing in other co-defendants. These co-defendants could consist of such people as the employer, the contractor, the subcontractor or professional consultants. When cases go to court this is often a long and costly process, and thus clauses in the contract attempt to determine who is liable for what, and who should bear the financial consequences.

10.5 Public liability insurance

The construction process is such that there are many situations where companies and individuals may be held liable for their actions. Therefore, it is crucial that these responsibilities be given due consideration and some form of insurance obtained. In the United Kingdom, construction companies are legally bound, by Clause 23 of the Compulsory Insurance Act 1969, to take out public liability insurance. Under these obligations, the contractor should insure against any damage, loss or injury to any property or to any person. In reality, most public liability insurance does not cover every eventuality since the underwriters of the insurance are unwilling to insure against an event such as accidental damage caused by a leak of radioactive fuel, or liability involving ships, aircraft or vessels. Furthermore, in an attempt to reduce premiums, events such as accidents caused by motor vehicles on the road will not be included, since these are covered by other types of insurance (e.g. motor insurance). In general, a good public liability insurance policy for a major construction project should include the following important points:

- The insurance covers all ways in which the insurer can be held legally liable, i.e. negligence, strict liability, nuisance, etc.
- The insurers are legally liable to pay any damages awarded against the insured.
- Awards indirectly relating to the injury must be covered, e.g. the hotel bills of a person whose house was negligently damaged by the insured.
- The policy protects against bodily injury to third parties including death, illness, accident, etc.
- The insured must be protected against liability arising from loss or damage to property. This may include theft of materials from a site.

Many events may be excluded for a number of reasons, for example, because a risk is covered by another insurance or a risk is uninsurable. Careful attention should be given to these exclusions, and the document should be read as a whole. In order for a policy to be valid, the insurers will place certain conditions upon the contractor which must be met before any payments will be made. These will be written into the policy and include:

- the damage or injury must be accidental and must arise in connection with the business as described in the policy;
- the injury occurred within the territories as specified by the policy, e.g. this could refer to one country only; and
- the injury, loss or damage occurred within the period of the policy. It should be noted here that policies known as run-off policies are available to cover liability arising up to six months after the completion of a contract.

10.6 Accidental injury to workers

The purpose of an employer's liability policy is to protect the contractor against any legal liability for accidents to, or diseases sustained by, the contractor's employees in the course of their employment. The employees covered are either those directly employed by the contractor or those employed by subcontractors. In the United Kingdom, the legal requirements relating to employees are specified in the Compulsory Insurance Act. The ICE Conditions of Contract (1969) have incorporated provisions in case the situation should arise where both the client and the contractor have contributed to the loss. The principal requirements are set out in Clauses 22, 23 and 24. The main thrust of these clauses is that the contractor must completely indemnify the client against any damages and compensation that may arise from the actions of workers injured during the contract. By law, the policy should cover £2 million per occurrence, but in reality the insurance is often unlimited.

A typical policy covers only the employees of a contractor, such as those under a contract of service or an apprenticeship. There is a "grey" area in the law as to whether or not self-employed contractors or labour gangs should be considered employees of the contractor. However, a good insurance policy would also underwrite them as employees, as the courts may well look below the surface in order to establish the true relationship. The contractor is protected against any claims for bodily injury or disease, and all legal and negotiation costs, as well as expenses incurred in settling the claim, should be covered by the policy.

As with all policies, certain conditions have to be met before any money is paid out. These may include the proviso that the injury or disease was caused during the period of the insurance (for example, an employee might contract asbestosis some 30 years after being exposed to asbestos, but if the exposure was at work and the employer had taken out insurance throughout the 30 years, then a claim could be made on the policy). In addition, the injury or disease must have arisen as a result of the business described in the policy; this means that a careful description of the business must always be given in the policy.

10.7 Professional indemnity insurance

The type of insurance policy discussed so far has been the cover that would be taken out by companies in the form of a public liability policy. However, this does not cover or guarantee the professional competence of the insured party. Claims against incompetence may result in financial loss rather than bodily injury, as previously discussed. Since around 1963, certain changes in English law have meant that liability can be attributed to a single person, especially in the region of professional negligence. Before 1963, the professional person was liable only to the client since it was the client who paid for the services. However, after the case of *Hedley Byrne* v. *Heller and Partners*, liability was extended to third parties, even though they were not in a contractual relationship with the professional.

As with public liability involving negligence against a construction company, the key question asked when determining if an individual professional has been negligent is: "Would other professionals have acted in the same way?" If the answer is "yes", then that person cannot be held liable. The American case of *Groover* v. *Simpson* illustrates the point well. A designer had been commissioned by a company to build a control room within a factory. When the room had been built, an employee was injured while working on a machine that was being controlled from the room. The employee sued for damages, and the court found that a poor control room design had resulted in the accident. Therefore, the employee was allowed not only to sue the employer for negligence, but also the designer.

The number of incidents in which individual designers have been held liable has increased dramatically over the past 20 years. In some countries, it has even come to the point of employers suing their own engineers for negligent design. A typical professional indemnity policy should include the following terms:

- Cover is provided only for claims made for a breach of "professional" duty.
- It applies only to claims made during the period of insurance and is known as a "claims made" policy. However, if a claim is made today for an error made 20 years ago, a current policy should provide the cover.
- The cover relates to negligence due to error or omission. In the United Kingdom, prior to 1985, it was possible to extend the policy to cover such things as public liability. However, owing to major losses by the underwriters, this type of cover has become increasingly difficult to obtain.
- There will be a maximum limit to the amount of cover given to the professional in any one year.
- The insurers will pay any legal costs in the defence of the claim. If these costs exceed the limit of the cover, a proportion of the additional costs will have to be paid by the insured.
- As with other types of insurance, the business of the professional must be carefully defined.

In addition to the above, insurance brokers will always produce a set of policy conditions that must be satisfied. The conditions vary from policy to policy, but a typical condition may be that the insured must in no way accept responsibility for a claim. Failure to comply with these conditions may result in the policy becoming void.

10.8 Additional policies for construction work

In addition to general third party liability insurance, due consideration should be given to policies that have been developed for the construction industry, such as Contractors' All Risks insurance, Erection All Risks insurance, Contract Works All Risks insurance and Insurance for Commercial Operations. These are discussed below.

10.8.1 Contractor's All Risks insurance (CAR)

Contractor's All Risks (CAR) insurance for building and civil engineering works is one of the newer types of technical insurance cover. It was first used in the United Kingdom to cover the construction of the Lambeth Bridge across the river Thames in 1929. Not long afterwards, a special CAR policy was developed in Germany and adopted elsewhere.

However, it was not until the construction boom just after the Second World War that real improvements were made.

Property underwriters were initially responsible for most CAR business and many would argue that CAR is not unlike property insurance. However, the number of CAR policies is usually relatively small and consequently the spread of risk is considerably less than in property insurance. The premium for most CAR policies is around 5 per cent of property premiums.

Although there are exclusions, CAR offers more protection against engineering and environmental risks than under traditional property insurance. Physical damage to project materials is insurable under the CAR policy. Not included are outbreaks of war, nuclear accidents, wilful negligence and similar events.

Although ancillary cover is available for third party insurance, additional policies must be taken out for certain types of risk. Purely financial exposure, such as non-completion, has to be insured through special cover or bank guarantees (see section 10.9). The CAR policy also excludes workers' compensation as there is usually a legal requirement for this type of cover.

10.8.2 Erection All Risks insurance (EAR)

In order to deal with the specific insurance needs of erection works, the Erection All Risks (EAR) policy was developed from the CAR policy. It covers the erection of machinery and structures involving little civil engineering work (e.g. foundations, machine beds, grouting). Most EAR policies have two sections: the materials damage section; and the third party liability section. The extent of its cover is fairly comprehensive, but the policy only forms part of the overall insurance required. In addition to general third party liability insurance, one of the following policies, as defined in the Swiss Reinsurance Company publication, *Introduction to Erection All Risks insurance* (1977), should be considered.

10.8.3 Contract Works All Risks insurance

Contract Works All Risk insurance policies combine the cover provided under CAR and EAR policies, but with a clear separation between the two types of contract work as far as sums insured, premiums and specific conditions are concerned.

10.8.4 Insurance for Commercial Operations

Insurance for Commercial Operations policies cover the risks of machinery breakdown, fines, loss of profits, and so on.

10.9 Advance Loss of Profits insurance (ALOP)

10.9.1 Background

Advance Loss of Profits (ALOP) insurance is referred to by many different names. These include Loss of Profits, Delayed Earnings Insurance, Delay in Start-Up Cover, Loss of Rent, and so on.

ALOP insurance dates back to the early 1930s, but the real development took place with the economic recovery following the industrial recession of the Second World War. The recession had led to a stagnation in the volume of premiums from the more traditional technical insurance lines. As a result, many brokers and companies ventured into new forms of insurance lines and in particular into any kind of consequential loss insurance. The main problem associated with this type of insurance is the wording of the policy terms, which are sometimes ambiguous resulting in disputes within the insurance industry. This led to the launch of a new scheme, Secured Protection Of Neglected Self-Assured Over-Run Risks (SPONSOR), designed to cover delays caused by:

- the failure of critical components to perform as expected;
- the failure of manufacturers, suppliers, contractors, etc., to meet contractual delivery or completion dates; and
- unforeseen legal restrictions.

The use of concepts such as BOT (Build-Operate-Transfer) and BOO (Build-Own-Operate) to stimulate industrial development in poor but rapidly developing countries has presented insurers with problems which are interesting but at the same time difficult to accommodate. The use of ALOP has yet to be fully mobilized by the brokers. There tends to be considerable interest in having ALOP insurance but, once negotiations start, the initial interest declines and many policies fail to materialize as a result of:

- the interests of the insured parties not being clearly specified and defined;
- high potential for disputes resulting from ambiguous policy wordings; and
- inadequate pricing.

10.9.2 Typical ALOP policies

Many underwriters are trying to use their experience to clarify ambiguities surrounding the wordings of ALOP policies with a view to developing and improving the ALOP issue. Such work requires careful attention, as even a single ALOP loss could have a devastating effect on an account, since the loss potential is considerable and the premium volume written is relatively small.

Most reinsurance companies, including the Swiss Reinsurance Company, have stated that they have overcome most of these problems through the improvement of the ALOP policy wording. Following a recent overhaul of its internal rating guidelines, the Swiss Reinsurance Company has also developed a more concise concept of ALOP cover, comprising the sections discussed below. In its opinion, this complex line of business can be made more manageable if the new ALOP strategy is adopted.

(a) The policy cover

The cover is usually restricted to the actual loss of gross profit resulting from an indemnifiable covered loss under a CAR or EAR policy culminating in a delay in the anticipated completion date of a project. However, the cover does not embrace the full extent of events covered under CAR or EAR policies (see exclusions).

(b) The insured party

Only the project sponsor or owner, as defined in the CAR or EAR policy, is covered by the policy.

(c) Exclusions

As specified earlier, the cover does not embrace the full extent of risk under CAR and EAR policies. Special exclusions for ALOP covers are provided whereby insurers shall not be liable for losses resulting from delay caused by any of the following:

- extensions or alterations of the specific exclusions of the material damage sections of EAR and/or EAR policies unless otherwise agreed in writing;
- restrictions by public authorities;
- the cost of any alterations, additions and/or improvements which may be undertaken after the occurrence of a material damage accident;
- earthquake, volcanic eruption, hurricane, flood or tsunami (tidal wave) unless otherwise agreed in writing;
- loss of or damage to the contractor's plant and equipment, loss or damage to existing property, or objects in the temporary care of the insured; and
- loss of contract, shortage of funds, riot, strike and penalties for non-completion, etc.

(d) Period of insurance

The erection, testing, commissioning and maintenance periods are usually all included. The length of the period of insurance is a function of the duration of contract works. The underwriter needs information regarding the nature of the testing phase in order to decide the starting dates of the testing and commissioning periods. The period of insurance is the same as the underlying CAR or EAR policy.

(e) Sum insured

The sum insured has to be defined for each policy and can be the expected annual gross profit, revenue, rent or fixed costs.

(f) Loss settlement

A loss is settled in line with the rules set out in the policy. There must be a prompt notification of the claim and, if the size of the claim justifies it, the site should be visited to obtain all the facts about the extent of loss or damage. The delay period, which forms the basis of indemnity, starts at the date when the project should have been completed had no accident occurred, but not earlier than the completion date as defined in the policy schedule. Indemnification is made on the basis of the actual incurred loss of gross profit or rent for the actual period of delay. However, such a period must not exceed the agreed period of indemnity.

10.9.3 Underwriting information

The information needed to underwrite ALOP is similar to that of EAR/CAR coverage, but varies from risk to risk. The insurance company must be able to assess the extent of exposure if it is to be fully aware of the risk it is taking. Basic requirements for underwriting this type of policy include the following:

- extent of property damage cover – copy of material damage policy;
- detailed works programme;
- list of major items and/or machines to be constructed/installed, with indication of relative importance, time for replacement, and spares;
- simplified process flow diagram;
- repair facilities available;
- information on the reputation and qualification of the contractors involved;
- period of indemnity;
- envisaged time excess or deductible; and
- sum insured, with details of calculation.

The lack of sufficient statistical information regarding the ALOP issue makes it difficult for underwriters to establish truly credible underwriting terms. Most policies are still priced on a trial and error basis. To overcome some of these problems, the basis of EAR or CAR can be used and adjusted as required; for example, in CAR-ALOP the material damage rate can be adjusted by various parameters.

10.10 Guarantees and bonds

There are several types of guarantee and bond that can be used to guarantee completion and performance as specified in the construction contract. These include:

- performance bonds;
- tender bonds;
- advance payment bonds;
- retention bonds; and
- maintenance bonds.

Performance bonds are dealt with in detail below, while the other types of bond are described briefly.

10.11 Performance bonds

In the current economic climate, contractors often get into financial difficulties and as a result are unable to complete a contract. Sometimes the situation of the contractor is so bad that the company is forced into liquidation and the client is unable to recoup the financial losses. These losses may be caused by subsequent delays in project completion or increased costs when a new contractor is appointed. To cover such situations, a contractor may take out a financial guarantee in the form of a performance bond.

Generally, a performance bond is an agreement between three parties: the client, the contractor and the person who issues the bond, known as the "surety". The role of the surety is to "bond" itself to the contractor, and in doing so it offers a guarantee to the client that the contractor will complete his or her obligation to finish the contract. If for some reason the contractor cannot complete the contract, then the surety steps in and pays the client a sum of money equal to the loss incurred.

Performance bonds should never be confused with insurance; they are completely separate for a variety of reasons. Insurance is normally an agreement between two parties (the insured and the insurer), whereas bonds involve three parties (the contractor, the client and the surety). The crucial point is that a bond offers no security to the contractor. With insurance, taking an act of negligence as an example, the contractor is effectively excused because, so long as adequate cover has been obtained, the insurance policy will pay for most of the legal costs and penalties. With a performance bond, the surety is not releasing the contractor from any responsibilities, but is merely there to assure the client that the work will be done. If for some reason the surety has to make payment to the client, it is entitled to claim the money back from the contractor through the courts or by some other method.

The type of bond used under the ICE Conditions of Contract has been developed to overcome the problems encountered with more traditional bonds under common law. For example, any subsequent changes to the contract or time extensions do not invalidate the bond.

A bond remains in effect until such a time as the contractor has completed his or her obligation. This could be for a considerable time after the completion date, since defects may not become evident straight away. However, most bonds will state an expiry date which corresponds to the end of the maintenance period and the issue of a maintenance certificate. Under ICE Conditions, the contractor and the surety are held and firmly bound to the client. This gives the client the right to sue either the contractor or the surety for damages.

Bonds of this sort often involve either banks or insurance companies as the surety. When a bank acts as a surety, it is often in the form of an extension of credit to a long-established customer. The implications of this are that the risks will probably have not been so well investigated as with an insurance company offering the same type of bond, because the insurance company will have undertaken a more thorough analysis of the contractor's financial stability. The main reason for selecting a bank as a surety in preference to an insurance company is usually that bonds from banks will be speedily dealt with. However, the contractor's ability to borrow further sums of money could be reduced. With an insurance company, it normally takes longer to process an application for a bond, but the employer can be assured that a contractor with a bond from an insurance company can be considered to be relatively stable.

In international transactions, surety bonds are an unconditional obligation to pay on demand a sum of money when the beneficiary considers that the contractor has failed to perform. In most cases, international surety bonds are issued by banks in the form of an unconditional letter of credit payable on demand without proof of non-performance. The American domestic bond market has several differences from the international market, the main ones being that the obligation is to perform the contract and the bonds can only be called upon by proof of default.

Finally, it should be remembered that the size of the bond is usually in the region of 10 per cent of the contract price. However, it is up to the client to decide if the size of the bond is sufficient.

10.12 Other forms of bond

Most other forms of bond are similar in style to performance bonds, but they are used to cover different aspects of construction, as indicated below:

- *Tender bonds*

  The purpose of a bid or tender bond is to ensure that the tender put in by the contractor is a serious bid. If the bid is accepted and the contractor fails to go through with it, the bond will cover any expenses incurred while the employer tries to find a new contractor.

- *Advance payment bonds*

  Advance payment bonds may be required to enable the contractor to purchase goods and equipment before construction has started. The bond protects the client against advances being lost through the contractor's default.

- *Retention bonds*

  Most construction projects will require money to be retained in order to provide a fund to rectify any mistakes made by the contractor. Retention bonds may be issued in lieu of these retention monies.

- *Maintenance bonds*

  Maintenance bonds can be used to provide the funds to correct defaults in construction that are discovered after the project has been completed. It is common for performance and retention bonds to be converted into maintenance bonds once construction has finished.

10.13 Summary

The preceding chapters have concentrated mainly on the various sources of finance available to the construction industry. In selecting which source of finance to pursue, the project promoter must consider many issues. For example, some sources of finance can be used only for specific purposes, while others can be used only for certain durations. Whichever source is selected the promoter must ensure that the cost of finance is low enough to guarantee that the financed project is successful and profitable. In order to keep costs low, it is essential that project risks are reduced. Throughout both the construction of a project and its subsequent operation, the invested capital will be at risk. Previous chapters have discussed some of the risk-sharing devices available (e.g. guarantees). However, this chapter provides an overview of important issues relating to insurance and performance bonds that must be considered as part of the overall investment package.

References

Ashley, D. B. 1980. "Insurance for major construction projects", in *Journal of the Construction Division* (New York, American Society of Civil Engineers), Vol. 106, No. 3, Sep., pp. 307-313.

Davies, M. C. (ed). 1989. *Avoiding claims. A practical guide to limiting liability in the construction industry* (London, E. & F. N. Spon), p. 256.

Eaglestone, F. N. 1979. *Insurance for the construction industry* (London, Godwin).

——. 1985. *Insurance under the ICE Contract* (London, Godwin).

Hinchley, J. W., 1986 "Payment and performance bond coverage and claims", in *Arbitration Journal* (New York), Vol. 41, No. 2, pp. 25-33.

Hinze, J; Abdul-Baki, W. 1981. "Insurance practices of utility contractors", in *Journal of the Construction Division*, Vol. 107, No. 3, Sep., pp. 413-423.

Insurance Institute of London, Advanced Study Group 208A. 1985. *Construction and erection insurance* (London).

Madge, P. 1990. *Civil engineering insurance and bonding* (London, Thomas Telford).

Mastrandrea, F. 1986. "The liability of the construction project manager", in *Construction Management and Economics* (London), Vol. 4, No. 2, Autumn, pp. 105, 134.

Reynolds, G. M. 1980. "Legal and insurance trends in engineering", in *Journal of the Engineering Issues Division* (American Society of Civil Engineers), Vol. 106, No. 1, Jan., pp. 3-10.

Rollage, M. P.; Neuschaefer, R. 1988. "Finding the right angle for working with a bonding company", in *Construction Plant Association Journal* (Pennsylvania), Vol. 59, No. 2, pp. 22-25.

Swiss Reinsurance Company. 1977. *Introduction to Erection All Risks insurance* (Zurich).

——. 1990. *Contractors All Risks insurance* (Zurich).

Varley, I. M. 1985. "Second Congress of Civil Engineers: The engineer in society 3. Liability, insurance and the engineer", in *Proceedings of the Institution of Civil Engineers* (London), Part 1, Vol. 78, June, pp. 569-571.

Vince, C. R. 1989. "Unified risk insurance: An update", in *Journal of Performance of Construction Facilities* (American Society of Civil Engineers), Vol. 3, No. 4, Nov., pp. 226-234.

White, G. W.; Mitcham, E. W. 1976. "Lessons to be learned from construction industry insurance claims", in *Structural Engineer* (London), Vol. 54, No. 2, Feb., pp. 53-58.

Wright, J. D. 1991a. *Construction insurance* (London, Chartered Insurance Institute).

——. 1991b. *Construction insurance (Insurance learner's series)* (London, Witherby).